FARMING WOMEN

Farming Women

Gender, Work and Family Enterprise

Sarah Whatmore
Lecturer in Geography
University of Bristol

First published 1991

Published by
MACMILLAN ACADEMIC AND PROFESSIONAL LTD
Houndmills, Basingstoke, Hampshire RG21 2XS
and London
Companies and representatives
throughout the world

Typeset by Footnote Graphics,
Warminster, Wiltshire

British Library Cataloguing in Publication Data
Whatmore, Sarah
Farming women: gender, work and family enterprise.
1. Agricultural industries. Family farms
I. Title
338.642
ISBN 978-1-349-11617-1   ISBN 978-1-349-11615-7 (eBook)
DOI 10.1007/978-1-349-11615-7

For my parents

Contents

List of Figures, Tables and Maps   ix

Preface and Acknowledgements   xi

List of Abbreviations   xiii

1 Introduction   **1**
Feminism and family enterprise   1
Women and farming   4
Concepts and themes   7
Structure of the book   10

2 Family Farming   **12**
A challenge to theory   12
External constraints   15
Internal resilience   19
Recent developments   24

3 A Feminist Reconstruction   **28**
The elusive family   28
Beyond the unity of capital and labour   30
Patriarchal gender relations   34
Domestic political economy   43

4 Theory into Practice   **46**
Making women's work count   46
Macro and micro analysis   49
Fieldwork and methods   51

5 Women's Work and Property   **65**
Patriarchy or life-cycle?   65
Gender divisions of family labour   66
Patriarchal labour relations   73
Patterns and problems   82

6 'Being the Farmer's Wife'   **86**
Ideology and the labour process   86
Family ties and working lives   87
Contrasting ideologies of wifehood   92
Processes and problems   102

7 The Domestic Political Economy of Six Family Farms   **105**
Two faces of family farming   106
The family farm in transition   117
Family farming under strain   127

8 Conclusions   **139**
Rethinking family farming   139
Shifting perspectives   144

Appendix: Survey Questionnaire   149

Notes and References   158

Bibliography   168

Index   184

List of Figures, Tables and Maps

Figures

2.1 The 'double specification' of agrarian petty commodity production 14
2.2 The position of farming in the 'agroindustrial complex' 17
2.3 Autonomous and market-dependent reproduction 23
3.1 Dualistic approaches to production and reproduction 38
3.2 Production, reproduction and livelihood 39
4.1 A relational typology of farm enterprises 55
4.2 Distribution of farms across the typology matrix, by study area 56
4.3 Distribution of responses to the survey of 'farm wives' by farm type for each study area 58
7.1 Household structure at Holly Farm 108
7.2 The labour process at Holly Farm 109.
7.3 Budget structure at Holly Farm 110
7.4 Household structure at Naylors Farm 112
7.5 The labour process at Naylors Farm 114
7.6 Budget structure at Naylors Farm 116
7.7 Household structure at Fountain Farm 119
7.8 The labour process at Fountain Farm 120
7.9 Budget structure at Fountain Farm 122
7.10 Household structure at Castleton 124
7.11 The labour process at Castleton 125
7.12 Budget structure at Castleton 127
7.13 Household structure at Vale Farm 129
7.14 The labour process at Vale Farm 130
7.15 Budget structure at Vale Farm 132
7.16 Household structure at Rough Farm 134
7.17 The labour process at Rough Farm 135
7.18 Budget structure at Rough Farm 137

Tables

4.1   Selected features of farms in Dorset and the
      Metropolitan Green Belt   54
4.2   The case studies   61
5.1   Women receiving resistance with domestic household
      labour   67
5.2   Women's agricultural labour   68
5.3   Women's agricultural labour, by farm labour type   70
5.4   Women's non-agricultural farm labour   72
5.5   Women's ownership of land, by farm type   75
5.6   Women's ownership of capital, by farm type   75
5.7   Women's independent income, by farm type   78
5.8   Women's ownership of capital in relation to their
      agricultural labour   79
5.9   Women's participation in decision-making, by farm
      type   80
5.10 Women receiving pay, by farm type   82

Map

4.1   Geographical location of the study areas   52

Preface and Acknowledgements

Farming Women presents a reassessment of family farming at a time when family enterprise is gaining in significance and interest beyond the farming arena. The book offers a feminist critique and reconstruction of petty commodity production, the concept widely used to describe various forms of small-scale production for the market based on family, or household, labour and property. Through a detailed study of family farming in England, the political economy of family-based production is examined as a unity of household and enterprise, intimately structured by patriarchal gender relations.

Placing women at the centre of analysis, the book challenges the prevailing invisibility of women in farming and family enterprise, and portrays their distinctive experiences as 'farm wives'. I am deeply indebted to all the women (and their families) who took part in the research. I should particularly like to thank those who put so much of their time and energy into the case studies. My representation of their lives is not their own, but I hope that it does justice to some of what they taught me.

This book began life as a doctoral thesis and, despite some rewriting, still bears the hallmarks of these origins. The research on which it is based was carried out on a part-time basis, between 1985 and 1988, while I was employed on a larger project on agricultural change funded by the Economic and Social Research Council and directed by Professor Richard Munton (University College London) and Dr Terry Marsden (Southbank Polytechnic). My debt to them in this piece of research, and much else besides, is enormous. The Department of Geography at UCL provided a stimulating workplace and I am grateful to many individuals within it for support of various kinds. In particular, I should like to thank Professor Ron Cooke and Claudette John for finding ways of supporting my work beyond the terms of any contract; Melanie Limb, who patiently introduced me to microcomputing; Jacqui Burgess and Sandra Wallman for valuable advice on methods; and Peter Wood for his constructive scepticism. Peter Jackson deserves special thanks for commenting on two versions of the full text.

A number of other people also made significant contributions to the production of this book which I should like to acknowledge. Ruth Gasson gave early and sustained encouragement to my endeavours. Martine Berlan, Juliette Caniou, Gema Canoves and Maria Dolors Garcia Ramon offered valuable comparative perspectives on farm women from their own research. Andrew Sayer made helpful comments on the typology, Mike Savage drew my attention to some useful parallels outside farming, and Michael Redclift and Doreen Massey provided important criticisms of the thesis version and encouragement in getting it published. Friends and colleagues in the Women and Geography Study Group (WGSG) provided much needed sustenance in the pursuit of feminist research. Tim Hadwin and John Dixon, at the School of Geography, University of Leeds redrew the figures with their usual skill and good humour.

Finally, my love and thanks go to Chris Whatmore and Jo Little, who shared a home and saw me through the trials of thesis writing; to Pickle (now sadly missed) and Spike, cats who gave joy even in the bleakest moments during that period; and to my parents for their unerring support. The book is dedicated to them.

Bristol   SARAH WHATMORE

List of Abbreviations

AEDC	Agricultural Economic Development Committee
BSA	British Sociological Association
CB	citizens' band radio
CCCS	Centre for Contemporary Cultural Studies
DCP	domestic commodity production
EEC	European Economic Community
EOC	Equal Opportunities Commission
ESRC	Economic and Social Research Council
GLC	Greater London Council
MGB	Metropolitan Green Belt
PCP	petty commodity production
PYO	pick-your-own (fruit and vegetables)
SCP	simple commodity production
WGSG	Women and Geography Study Group

1 Introduction

The recent rediscovery of a world of economic activity 'outside' the institutions of corporate capitalism has upset their widely accepted status as a definitive feature of modern life. This rediscovery was triggered initially by evidence of the growth of household-based production in developing countries undergoing capitalist transformation in the 1960s and 1970s (Bromley and Gerry, 1979). More recently it has been informed by the experience of economic restructuring in advanced industrial countries in the 1980s. Under conditions of global recession, mass unemployment and of widespread state deregulation a variety of forms of non-wage work and non-corporate production, based on family, household and community ties, have re-emerged as socially and economically significant in western European countries (Mingione, 1983).

The convergence of these trends and experiences is reflected in a growing dialogue between third- and first-world research ideas and agendas associated with an overriding concern to reconnect the structures and practices of consumption, at the level of the household, with those of production in contemporary capitalism (Redclift and Mingione, 1985).[1] A common focus has been the ways in which people make a living, bringing to light the diversity of livelihood strategies and work practices which exist under capitalism but which have been eclipsed by mainstream and Marxist preoccupations with corporate capital and wage labour. However the plethora of ill-defined and poorly related terms used to describe these strategies and practices, centring on notions of 'informality' or 'domesticity', indicates the difficulties of constructing a coherent theoretical analysis of such phenomena which appear anomalous in terms of orthodox theories of the development of industrial capitalism.

The Marxist concept of petty commodity production (PCP) has been looked to as a response to these theoretical problems. PCP defines a variety of types of small-scale production for the market based on family, or household, labour and property. In Marxist terms PCP represents a distinctive form of production because it contradicts the tendency towards the separation of capital and labour. As such, the place of PCP within capitalism – its features, dynamics, historical

1

fortunes and political implications – have been a long-standing interest in Marxist political economy (Bernstein, 1986, p. 9).

This interest can be traced through studies of the historical role of family enterprise in the development of early industrial capitalism (Davidoff and Hall, 1987) and the contemporary role of household production in developing countries (Long, 1984). However, until now, the significance of PCP in modern industrial countries has received much less attention (Bechhofer and Elliott, 1981) on the assumption that it represents an inherently transitional form of production only apparent in societies at 'earlier stages' of capitalist development. An important exception to this general neglect has been agriculture, which is the only major sector of industrial production in advanced capitalist societies still dominated at the farm level by family enterprise. As a consequence, theoretical debates about the persistence of the family farm are now emerging from their marginalised status as a 'special case', to be hailed as pertinent and instructive to the analysis of these 'new' livelihood strategies in advanced societies at large (Williams, 1984; Benvenuti, 1985).

Yet the concepts of orthodox Marxism, including PCP, present major difficulties for the analysis of family enterprise and household production because they are founded upon an opposition between political economy and domestic economy which such forms of production confound. They are embedded in a dualistic conceptual framework which counterposes family and economy, reproduction and production and assumes that, under capitalism, home and work represent two functionally and spatially separate domains. In consequence these domains have tended to be studied in isolation (Rowbotham, 1972). Where the domain of the economy has been elevated to a primary, or even sole, focus of study (see Urry, 1981), that of the family has received limited attention. Instead there exists a pervasive, but often implicit idea of *the* family, as the nuclear family, and of its functional relation to the wage-economy (Close and Collins, 1985).

Feminist scholarship has played a significant role in challenging such assumptions and in the rediscovery of non-wage work in general (Moore, 1988). Through its primary concern to make sense of women's lives it has brought into focus those aspects of the social and economic world – family, reproduction and the home – hidden by rival perspectives. As critique, much of its effort has been directed at challenging the tenets of Marxist analysis. In particular, through the exploration of women's experience as domestic workers, it has effectively challenged orthodox concepts of labour and exposed the family as an important site of exploitation and struggle (Barrett, 1980).

While these undoubtedly represent important and widely influential feminist ideas, much feminist analysis and, specifically, the key concept of patriarchy remain closely bound up in the dualistic framework which it had sought to challenge. In particular, the treatment of women's 'domestic labour' and of the processes of reproduction in general as part of a discrete domestic realm – the site of patriarchy – remains bound to this pattern of thinking, with its roots in the powerful spatial imagery of home and work (WGSG, 1984). This kind of analysis of the labour process, and its role in sustaining patriarchal gender relations, becomes tautologous. As Redclift has argued, it is unclear whether 'the processes of reproduction [are] defined by the fact that they occur in a particular place, "the home", or [if] our idea of what constitutes the domestic [is] defined as the site of certain specific processes' (1985, p. 97).

Debates within feminism have already criticised this approach as ethnocentric (Kazi *et al.*, 1986) and class biased (D. Smith, 1983), uninformed by the experiences of black women in western societies and women in non-western societies. It is an approach which has also meant that, while there are some fine studies of women as waged workers (Cockburn, 1985; Pringle, 1988) and women as wives (Finch, 1983; Callan and Ardener, 1984), feminism has had very little to say about women's involvement in PCP, where their position cannot readily be understood through the orthodox categories of either housewife or wage labourer. Such women are invisible as workers in family enterprises precisely because of the overriding ideology that they are really housewives who, as Moore has put it, 'happen to be using their leisure time in a profitable way' (1988, p. 84).

Farming Women offers a feminist critique and reconstruction of PCP, based on a detailed analysis of the particular case of family farming in contemporary Britain. It constructs a very different reading of women and family farming from the rustic romanticism of the 'farmer's wife' and the 'Home Farm' as popularly imagined. But these very images also lead us to the heart of the challenge which family enterprise presents to feminist analysis because here home and work, and all the cemented associations that these small words have come to signify, are located in the same place. How is it that the activities, indeed the existence, of women in such enterprises have nowhere informed the theories and concepts devised to analyse this form of production? What kinds of conceptual reconstruction are necessary to make sense of women's lives as family workers and the fundamentally gendered nature of PCP? In addressing these questions, the book explores the ideologies and practices of the family labour process as these structure the beguiling unity of the family farm and shape women's experience as wives within it.

WOMEN AND FARMING

Despite the far-reaching restructuring of agriculture in the postwar period, family farms remain the principal unit of agricultural production in advanced capitalist societies. In Britain, for example, there has been a marked concentration of agricultural production onto a smaller number of larger farm holdings (Harrison, 1982). Large farm units, defined as occupying more than 300 acres, have risen from less than 5 per cent of the total number of holdings and total farmed acreage in England and Wales in 1951 to 14 per cent and 54 per cent respectively by 1984 (Grigg, 1987). Even so, nearly three-quarters of farm holdings currently employ no full-time hired workers (Burrell *et al.*, 1984) while family labour makes up some 63 per cent of the total agricultural workforce in England and Wales (Errington, 1987), and in both cases the trend is upwards. Accurate figures on asset ownership are not available, but it is estimated that about 85 per cent of land and 95 per cent of business capital employed in the farming industry are family as opposed to corporately owned (Harrison, 1983).[2]

However the world of farming, like many other spheres of activity, has traditionally been depicted as a 'man's world' (Williams, 1964, p. 95). Analytical attention has focused on the 'farmer' as business principal, labourer and decision-maker (Errington *et al.*, 1986), the term itself carrying masculine connotations.[3] As a consequence the composite social character of the family farm has all too readily slipped from view. In particular, until the late 1970s, the role of women on the farm had received scant attention, on the implicit assumption that it was much the same as that of any other married woman, namely that '... after marriage men are gainfully employed while women work in the house and have children' (Littlejohn, 1963, p. 69).

The major contribution of the growing body of research on farm women in Europe and North America has been to show the inaccuracy of such an assumption and to catalogue the varied combinations of labour roles performed by women on the family farm. These include agricultural labourer, business partner, off-farm income earner and farm secretary.[4] Much of this work has been directed towards filling the vacuum left by official statistics on farm labour which tend to deal generically with the category 'family labour' and fail to distinguish individual members' contributions within this. In Britain, for example, the contribution of 'farm wives' to family labour has only been identified since 1977 (Errington, 1983). However, while such studies have demonstrated that women too are 'gainfully employed' on the farm, women's work 'in the house and with children' remains largely excluded from the activities considered relevant to an analysis of the family labour process (recent exceptions include Ghorayshi, 1989). Instead this area of women's work is taken as given and then offered as an explanation of the different pattern of involvement in agricultural, or other income-generating, work between men and women.

More recent research, informed by pioneering feminist work in the third world,[5] has argued that this narrow concept of work *mis*represents the farm labour process and *under*-represents women's participation in it (Reimer, 1986). It has highlighted the analytical constraints imposed by a concept of farm labour which counts as relevant only activities in the commercial agricultural production process and ignores a whole realm of conventionally defined 'women's work' in the subsistence and reproduction process. Such a concept obscures the essential *interdependence* between these two processes and of the labour relations which underlie them.

Furthermore traditional analyses ignore the fact that women are primarily involved in farming through specific forms of familial gender relations, most significantly through marriage, as wives, but also as the daughters (see Gasson, 1987) and mothers of men 'farmers'. In Britain, it is variously estimated that between 80 and 90 per cent of farmers are married men, while only between 1 and 3 per cent are women farming in their own right (AEDC, 1972).[6] In the literature, however, a distinction between women and 'wives' is rarely made and, more importantly, the significance of the distinction is almost entirely missed.[7] In general, farm women have been treated as a research category in a way which renders gender – the social process and experience of being a woman or a man – irrelevant to an *explanation* of their position. At worst, they are reduced to the analytical status of a 'factor of production' in a 'male' production process – land, labour, capital and wife (even if they are increasingly being seen as an 'undervalued asset' (Gasson, 1980).

Farming Women

There is a manifest failure in the literature on family farming, including that on farm women, to theorise gender relations as a principal axis of social division within the farm family which shapes men's and women's labour roles and identities as 'farmers' and 'farmer's wives'. Instead, the gender division of labour

> is either regarded as merely a technical division of labour which distributes agents into socially equivalent places in production, and/ or that it is a natural development which distributes agents on the basis of certain gender specific attributes. (Molyneux, 1977, p. 62)

This position is predicated on a widely held, but rarely examined, conception of 'the family' as a universal social institution structured around a natural or normative division of roles between men and women. It is a premise which can be traced through the business model of the family farm which prevails in agricultural economics (Gasson *et al.*, 1988), and the traditional structural–functionalist paradigm in rural sociology (Crow, 1985). Both traditions have been widely criticised on other grounds for their theoretical inadequacies (Newby, 1982).

They find a parallel, however, in the more robust theoretical tradition of Marxist political economy. Here the 'problem' of the family farm is seen in terms of its resistance to the process of capitalist transformation. A principal analytical distinction is drawn *between* family and capitalist producers, taking the presence on the farm of waged labour, as distinct from family labour, as the key indicator of transformation. The concept of the family and the social mechanisms of labour expropriation *within* the category family labour are left largely unexamined. Marx's term, 'natural economy', to describe pre-capitalist forms of domestic commodity production, is revealing in this respect (Folbre, 1982).

An analysis of women's *labour* as farm wives brings to the fore those dimensions of the farm labour process previously neglected by the narrow focus on agricultural production, and those mechanisms of labour expropriation within the family household previously ignored as a result of the preoccupation with the family/wage labour dichotomy. The reproduction process and the familial gender division of labour, together with the position of women as wives, constitute the hidden agenda in the political economy of petty commodity production.

CONCEPTS AND THEMES

This study centres on the family farm, taken to embrace a range of actual forms of production which combine class and gender relations in a variety of specific ways associated with different degrees of commoditisation. 'Commoditisation' describes the process by which the family household and farm enterprise are tied into the wider market economy in such a way that their form and conditions of existence are increasingly structured by it. The central question is thus cast not in terms of understanding the *survival* of the family farm as some kind of resilient pre-capitalist relic, but of understanding its *transformation* as a generically capitalist form of production. The analysis focuses on the transformation of family labour relations in the commoditisation process and, specifically, on the changing position of women as wives in the gender division of labour on the farm. Women's own articulation of their experience as farm wives is used to examine the contradictory, rather than simply functional, way in which familial gender divisions and ideologies are taken up and reshaped in the commoditisation process.

The approach adopted here remains committed to the central project of Marxist political economy, as a materialist framework of analysis concerned with the social division of the means of human subsistence and the products of human labour. However agrarian political economy still bears the heavy imprint of the structuralist brand of Marxism, an approach in which '. . . structures [are] credited with the power to develop in accordance with their own laws' (Bourdieu, 1977, p. 84). The framework and analysis presented here engage with feminist and post-structuralist developments in social theory which demand a radical rethinking of the meaning and practice of political economy.

Thus the commoditisation process remains an important focus as the main impetus for change in the structure of family farming and the wider agricultural industry. However, it is conceived of as a *social* process whose development, as Massey has argued in the context of industrial restructuring, 'is not a mechanistic outcome of external forces, whether conceived of as abstract immanent tendencies of a mode of production or specific historical processes' (1984, p. 15). Rather, it is understood as rooted in human agency and, by implication, human consciousness and subjective meaning (Thrift, 1983). Moreover the analysis recognises the causal powers of social structures and processes other than class and capital accumulation in shaping PCP and which cannot be reduced to them, however hard it is to disentangle their effects on the ground. In particular, gender relations are explored as a general category of social relations, which can take a variety of specific forms in an active process of conflict and struggle (Foord *et al.*, 1986).

The two elements making up the composite concept of family labour are re-examined. The farm labour process is first redefined to encompass productive and reproductive activities sustaining the farm household on the land, in which agricultural commodity production is but one, important, element. The farm labour process is analysed more broadly than in the existing political economy literature in two other senses also. Firstly, to encompass the different *relations* under which men and women work in terms of their control over the land and capital they work with, and over the income and goods which are its product. Secondly, to incorporate the ideological and experiential dimension of the labour process. The focus here is the *gender ideologies* and identities mobilised in the everyday work practices of farm households, which legitimise and mask the exploitative nature of the family enterprise. It is argued that this dimension of the labour process plays a crucial role in the construction and transformation of PCP.[8]

The highly chaotic concept of 'the family' is then broken down into its constituent elements of kinship, household and familial ideology (McIntosh, 1979). It is re-examined in terms of the *conjugal family household* as a specific historical and cultural combination of these elements which characterises the contemporary structure of the family farm in Britain and other advanced industrial countries.[9] The internal social divisions of this form of household change through the life-course of the married couple at its core, but are principally structured by gender relations and secondarily by generational seniority. Marriage is examined as a patriarchal institution, central to the reproduction of the property and labour relations characteristic of the family enterprise, through which men are empowered and women subordinated in specific ways (Barker and Allen, 1976; Young *et al.*, 1981).

Clearly the gender division of labour on the farm extends beyond the farm household to hired labour and, potentially, to an extended kin network of the resident family. Nor is it divorced from the pattern and structure of gender relations in wider society. However attention is restricted here to the particular position of women as wives, recognising that it is marriage which shapes most women's participation in the farm labour process. It will be argued that understanding what it means for women to be socially identified as a 'farmer's wife', whatever their labour role or property status on the farm, is crucial to an understanding of how their subordination in the family enterprise is structured and perpetuated. However, in adopting this focus, it is recognised that

> As wives, women are already undergoing the silencing or under-recognition of the rest of their personhood which allows them to be so designated. An analytical acquiescence in this process cannot be avoided and is indeed a necessary condition of understanding the forces which create and sustain it. (Callan and Ardener, 1984, p. 2)

A concept of *domestic political economy* is developed as a basis for analysing the intersection of gender and class relations in the transformation of PCP. The detailed analysis of the social divisions structuring the family farm in terms of both commercial enterprise and household livelihood represents an application of these principles to the specific circumstances of contemporary family farming in southern England. A distinction is made throughout the book between the general concepts being put forward for the analysis of PCP and the theoretical arguments specific to the circumstances of farming and the English case.

The book raises many themes and issues which it does not directly address, let alone resolve. It seems appropriate at this point to qualify the reader's expectations by clarifying some important limitations on the scope of the analysis. No attempt has been made to produce an historical reconstruction of the changing position of women as wives in the family farm. This is the focus of other research in Britain (Bouquet, 1986; Davidoff, 1986) and the United States (Sachs, 1983). The concern here is with the contemporary restructuring process. Empirically, family farms and farm women's experience are examined at one point in time, although clearly encompassing farms at different levels of commoditisation and women at different points in their life-course. The historical interpretation implicit in the analysis is that neither commoditisation nor the life-course can be understood as logical processes, with fixed stages to be passed through, only as lived processes, with constrained courses of action to be taken.

Similarly, while the analysis draws on field-research conducted in two different study areas, these are the 'locus' rather than the 'focus' of study. They represent a means of examining the specificity of family labour relations in different types of farming system and local conditions (Newby, 1985). While the recent vogue for 'locality studies' has been resisted, a number of geographical themes run through the analysis.[10] The two most persistent of these are the spatial configuration of productive and reproductive processes, and the importance of spatial metaphors (such as spheres, domains, arenas) in structuring our patterns of thinking about them; and the ways in which space is actively taken up and reshaped within the structuring and contestation of social relations, such as gender.

Finally, while it is recognised that the social relations of childhood and race are important dimensions of family farming, they are not given explicit treatment here.[11]

STRUCTURE OF THE BOOK

The book divides roughly into two parts. The first part explores the theoretical concepts and arguments related to the analysis of family farming as a form of PCP. In Chapter 2, current debates about family farming within agrarian political economy are examined to discover how the family and gender relations disappear from their field of vision. It concludes by identifying some important insights to be drawn from this literature, informed by recent criticisms and new departures within it. In Chapter 3 the limitations of Marxist social theory for the analysis of 'the family' and gender relations, and hence for understanding the family labour process, are developed. Drawing on feminist theory and research, the general concepts necessary to provide political economy with a viable means of addressing these questions are outlined. The final part of the chapter presents a revised political economy approach which is developed as a theoretical framework pertinent to the analysis of the particular case of family farming in Britain.

Chapter 4 provides the link between the conceptual and theoretical issues examined in the previous chapters and the empirical analysis presented in the second part of the book. It does so by drawing out the implications of the foregoing arguments for doing research on women's work. The second part of the chapter introduces the places and people at the heart of the study and outlines key features of the survey and ethnographic methods adopted. Technical details can be found in the appendix.

Chapter 5 begins the empirical study by sketching the broad

dimensions of the gender division of family labour on the farm, concentrating on the activities and conditions of women as farm wives. It argues that it is patriarchy rather than life-cycle which lies at the heart of these divisions, while the level of commoditisation of the farm makes an important difference to women's position and experience. In Chapter 6 women's own evaluations of their position as wives in the farm labour process are brought to the fore. Their ways of making sense of their experiences are examined in terms of various ideologies of 'wifehood' which are mobilised in the labour practices of farm families. The analysis highlights points of conflict arising *within* these ideologies and practices for women at a personal level, and *between* them and the process of commoditisation. The empirical analysis is brought to a close in Chapter 7 by broadening the focus from the specificity of women's experience of work to the ways in which gender acts as a fundamental structuring process in the wider *domestic political economy* of six case-study farms.

Chapter 8 draws together some conclusions from the analysis, in two parts. The first examines its implications for understanding the nature and transformation of contemporary family farming. The second returns to the more general issues raised in this introduction. It considers the wider significance of this research for theorising the boundaries and contours of gender and class, patriarchal and capitalist relations and for the future meaning and practice of political economy analysis.

2 Family Farming

A CHALLENGE TO THEORY

In combining family property ownership and family labour in commercial agricultural production, family farming represents a distinctive form of production in relation to the dominant features of modern industry, both in terms of the labour process and of the organisation of capital. In consequence it has attracted considerable research attention as 'a challenge to theory' (Friedmann, 1981, p. 10) because its persistence is hard to reconcile with prevalent theories of the process of capitalist development whose principal referents lie in the structures and dynamics of urban industrial capitalism (Buttel, 1982).

Marxist political economy has most influenced the terms in which this challenge has been taken up (Bradley, 1981).[1] This reflects the strength of agrarian political economy as a strand of Marxist analysis which can be traced from the legacy of the late nineteenth- and early twentieth-century classic Marxist texts on the transformation of the peasantry (Hussein and Tribe, 1983) through more recent developments in this same debate in the contempoary 'third world' (Goodman and Redclift, 1981). Redclift defines political economy as an approach which seeks to

> . . . locate economic analysis within specific social formations and explains the [restructuring] process in terms of its costs and benefits for different social classes. It recognises the specificity of social formations but seeks to explain structural variations within a coherent interpretive framework. (1984, p. 5)

This chapter looks at the specific discourse within this tradition concerned with the role of PCP in contemporary agriculture – a discourse which has been variously termed 'the decomposition question' (Buttel, 1982) or, more recently, 'the commoditisation debate' (Long, 1986). It explores how the terms of this discourse effectively eclipse the significance of 'the family', as a composite social grouping, and gender, as a primary division structuring the family labour process. The central concern in this discourse has been to explain the persistence of 'petty' or 'simple' commodity forms of agricultural production under capitalism against the transitional status ascribed them in more evolutionary interpretations of Marx's work.[2] It is a discourse which has been conducted in highly abstract terms in pursuit of the 'laws of motion' of the capitalist development of agriculture. Despite a gradual departure from this orthodox evolutionary line of thinking its legacy remains a powerful influence in current work.

By the mid-1980s, this 'debate' had reached something of an impasse, dominated by two rival explanations for the persistence of agrarian petty commodity production. These explanations are couched

> ... either in terms of the way in which capitalism selectively sustains certain forms of small-scale peasant or simple commodity production, which cheapen the reproduction of labour for the capitalist sector ... or in terms of a certain internal dynamic which generates social and cultural resistance to capitalism itself. (Long, 1984, p. 4)

The first of these interpretations can be characterised as the *constraints thesis*. Broadly, it is argued that the persistence of the family farm is less the result of any inherent robustness in PCP itself than of the constraints placed upon the development of capitalist relations of production on the farm by the biological nature of the agricultural production process. The petty commodity producer is conceived of as an 'outworker' of capital, exploited through the indirect appropriation of surplus value from the family labour process, either by means of devalorised labour time (for example, de Janvry, 1980a; Davis, 1980) or by means of unequal exchange (for example, Sinclair, 1980; Conway, 1981). Goodman and Redclift (1985, 1988, 1989) have provided the most recent and most sophisticated version of this thesis in the British literature.

The second interpretation can be characterised as the *resilience thesis*, which argues, broadly, that it is the integrity of the PCP labour process, drawing on internal reserves of family labour whose consumption is not mediated by a wage, which has ensured its survival. This approach builds on the early work of Kautsky (Banajii, 1976) and Chayanov (1966). In its application to contemporary advanced capitalist societies this thesis has been developed most fully in the pioneering work of Friedmann (1978a, 1981, 1986b), which has been highly influential internationally on research in this field, including Britain (see, for example, Winter, 1982, 1984; Bouquet, 1982, 1984b).

Both approaches are founded upon a 'double specification' (Fried-mann, 1978a) of domestic forms of production, as the intersection of processes external and internal to the unit of production (Winter, 1986, p. 62). These processes refer respectively to relations between PCP and the wider market economy and to those inside PCP itself defined by its dual status as both enterprise and household. This conceptualisation is illustrated in Figure 2.1. However the constraints and resilience theses focus on opposite components in this specifica-tion, as crystallised in a recent exchange between Goodman and Redclift, and Friedmann (*Sociologia Ruralis*, vols 25/26, 1985/6). This exchange provides a useful means to explore the wider debate. Each thesis is examined in turn in terms of its potential contribution to an integrated analysis. For, as Friedmann herself has suggested, '[they] are not alternatives but complementary dimensions of a com-plete analysis of the family farm in advanced capitalism' (1986a, p. 1).

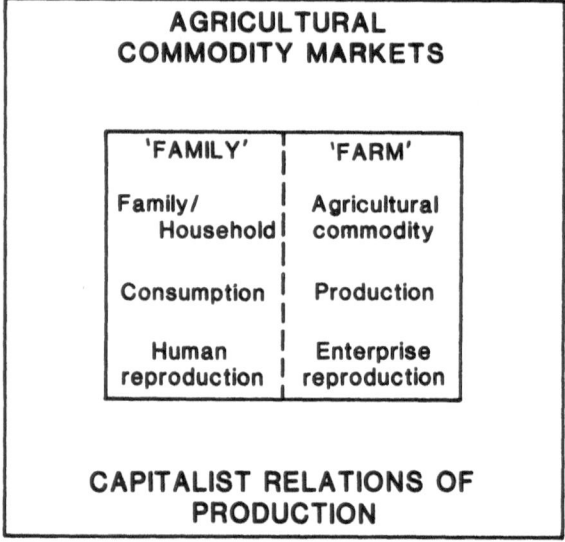

Figure 2.1   The 'double specification' of agrarian petty commodity production

In practice, however, this discursive division has tended to frag-ment rather than integrate the analysis of the internal and external relations of the family farm. Moves towards a more integrated approach have begun to emerge in the literature associated with the incorporation of 'post-structuralist' developments in Marxist social theory (Long *et al.*, 1986; Scott *et al.*, 1986). These developments are evaluated in the last part of the chapter as important steps forward in theorising PCP, but still unsatisfactory in their treatment of the *internal* relations of the family farm. In conclusion, the main points to be carried forward into the analytical framework developed in Chapter 3 are drawn together.

EXTERNAL CONSTRAINTS

The analysis of agrarian PCP was long dominated by a narrow sociological concern to determine the class status of farmers by distinguishing between *capitalist* and *family*, that is non-capitalist, forms of production. In this concern, Marxist research has shared a common problematic with earlier non-Marxist political economy traditions.[3] The strength of Goodman and Redclift's work is that it supersedes this preoccupation with formal categories. Their work focuses on a theoretical examination of the *process* of transformation in agricultural production relations, broadening the scope of analysis beyond the farm itself to encompass its changing position within a wider conception of agricultural production. They identify a quite different point of departure for agrarian political economy from its traditional treatment as a 'special case'. They argue that

agriculture [should be] considered much as any other sector in that [in the process of competition] industrial capitals seek to valorise all productive activities and to fully exploit technological innovations to maximise surplus value. (1985, p. 12)

Their analysis draws on Marx's concept of subsumption as the tendential process by which pre-capitalist relations of production are transformed or subordinated by capital. Following Marx, they distinguish between two types of subsumption. Formal subsumption refers to the process whereby capital subordinates an existing labour process without altering the technical means or the social relations of production. Here surplus value is extracted *indirectly* by such mechanisms as unequal exchange, devalorised labour time and credit relations (see Winter, 1984). It is the 'mode of compulsion' on the producer, rather than the internal relations of production, which is changed (Goodman and Redclift, 1985, p. 13). The second type, *real subsumption*, describes the capitalist transformation of an inherited labour process by revolutionising the technical means of production, requiring more complex divisions of labour and larger-scale production. Surplus value is here extracted directly as profit from the wage labour process. The crux of the argument is that the subsumption process operates in a distinctive way in agriculture. This is a result of the structural constraints imposed on the production process by nature 'as the biological conversion of energy, as biological time in plant growth and animal gestation and as space in land-based rural activities' (Goodman *et al.*, 1987, p. 2).

Within these constraints it is land, as an agricultural means of production, which has received most attention, particularly through the re-examination of Marx's theory of rent.[4] Three main points about the peculiar position of land in agricultural production can be summarily drawn from this work. Firstly, the agricultural use-value of land (its fertility in a broad sense) varies markedly across space, and is neither mobile nor reproducible (Ball, 1979; Murray, 1977). Secondly, the role of land as an organic means of production places temporal constraints upon the ability of capital to modify 'production time' as against labour time in the agricultural production process (Mann and Dickenson, 1978; Mooney, 1982). Thirdly, in agriculture land absorbs a substantial part of the capital invested in it, so that fertilisers, drainage and so on become an integral part of the land itself (Fine, 1978). These features influence the agricultural accumulation process directly, and indirectly through the medium of private landownership (Whatmore, 1986).

The development of agriculture under capitalism can be seen as a struggle to break the dependence of agriculture on the land as an organic means of production and to circumvent the bar of exclusive property rights over land by restructuring the labour process. The relation between capital and land is thus a dynamic one. As Marx suggests, 'fertility, although an objective property of the soil, always implies an economic relation to the existing chemical and mechanical level of development in agriculture' (1976, 3, p. 651). This point is developed by Goodman and Redclift in their analysis. They argue that the form and extent to which land and organic nature present obstacles to the uniform *real* subsumption of the agricultural labour process represents 'the objective limits of [capital's] capacity to transform the production process through the technological means of production at any one point in time' (1985, p. 18).

Within this context, the capitalist restructuring of agriculture is argued to have taken two main directions:

(i) The valorisation of agricultural products outside the land, or farm-based production process, at the input and output manufacturing stages of production.
(ii) In farming itself, through the replacement of land, or the modification of its role in the labour process, through the development of the technological means of production.

Modern agriculture, conceived of as the process of producing food and fibre, thus encompasses more than just farming and has become increasingly differentiated under advanced capitalism (Goss *et al.*, 1980). Farming itself has become merely the site for raising plants and animals within a much larger *agroindustrial complex*. This is illustrated in Figure 2.2. The production of technological inputs and

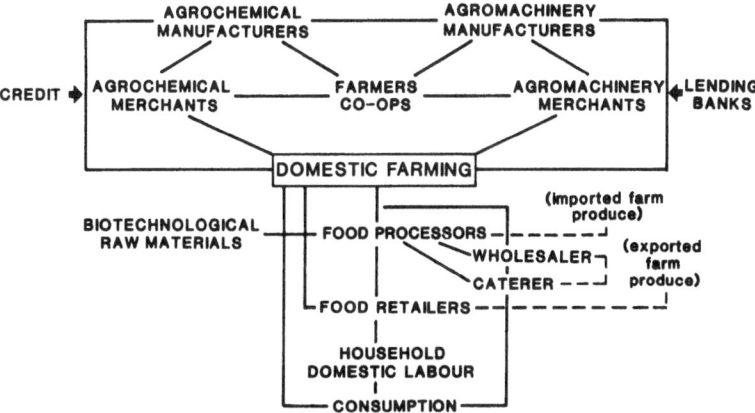

Figure 2.2 The position of farming in the 'agroindustrial complex'

the processing and marketing of farm products as food and other organic materials have been fully commoditised and taken off the farm in the processes of 'appropriationism' and 'substitutionism' (Goodman *et al.*, 1987).[5] The central role of the state in promoting these developments through various direct and indirect subsidies to technological development and adoption is a neglected aspect of the restructuring process in agriculture (but see Kloppenburg, 1988).

As Goodman and Redclift have argued

activities related to production and processing which at earlier conjunctures were regarded as integral elements of the land-based

production process [have been] progressively appropriated by
industrial capital. (1985, p. 16)

Both the agro-inputs and food processing sectors of the agroindustrial
complex are divorced from the land as a means of production and are
characterised by monopoly capital relations. Moreover these sectors
now account for a higher proportion of value added and product price
than farming itself (Newby and Utting, 1983; Friedland, 1982). It is in
this sense that the real subsumption of agriculture is not to be
observed on the farm itself but rather it is represented, so Goodman
and Redclift argue, 'in the tendential movement of capital to elimin-
ate the labour process as a land-based activity' (1985, p. 18).

The second direction taken in the agricultural restructuring process
has been the transformation of the *internal* relations of farm production. In a very few sectors of agriculture, land has been successfully
replaced as an essential means of production, permitting the full
subsumption of the labour process characterised by corporate ownership and management structures, large-scale production units and
wage labour relations.[6] None the less the limits of this kind of
revolutionary transformation of farm production, discussed above,
have meant that farming has been more commonly subordinated
indirectly.

Mechanisms of 'indirect' subsumption are associated with the
development of *external* relations, such as technological, credit and
contract sales relations, between the farm and industrial and finance
capitals. Through these relations surplus value is appropriated from
the farm labour process while the labour process itself is tied into
a wider system of accumulation, so that the internal relations of
production on the farm cannot be reproduced outside the reach of
capital (van der Ploeg, 1986, p. 52). These relations have been
termed elsewhere the 'technological treadmill' (see Cochrane, 1958).
In consequence, whilst the ownership of farm assets commonly
remains with the farm family, the farm family's control over the
management of the production process becomes compromised and
integrated within an externally determined dynamic, whether or not
the farm production process is characterised by family labour or wage
labour relations (Goodman and Redclift, 1985, p. 20).

There are a number of difficulties with the 'constraints thesis' and
with Goodman and Redclift's version of it. While in many ways
insightful as to the specificities of the agricultural production process,
their thesis remains wedded to a structuralist and economistic

conception of how that process works. The term subsumption, like decomposition, incorporates mechanistic assumptions about the direction of change. It is the logic of capital accumulation outside farming which is the irresistible 'motor' of restructuring riding roughshod over the activities and circumstances of social actors in the farm production process. In short, there is no conception of restructuring as an active social process, nor of any social structures or relations which may run parallel, or even counter, to capital penetration.

Perhaps the most serious problem is the treatment of the family farm itself. Because they focus on the changing *position* of the family farm within agricultural restructuring more widely, its internal structure remains largely unexamined. Despite their suggestions that the status of the family farm may form 'a continuum of intermediate states between the "pure" form [simple commodity producer] and the archetypal capitalist enterprise' (1985, p. 10), nowhere does the family farm become an *active* constituent in Goodman and Redclift's analysis. The contention of the 'resilience thesis' that the survival of family farms has something to do with the nature of their internal relations as forms of PCP is too readily dismissed by them as a product of 'the ideological nature of most of the discussion of family farming' (1985, p. 20). While the importance of populist ideologies to the terms of much of the debate is undeniable (see, for example, Sinclair, 1980; de Janvry, 1980b), the analytical perspective of the 'resilience thesis' and, more specifically, of Friedmann's work, is valuable in so far as it does throw some light on the dark interior of the family farm.

INTERNAL RESILIENCE

Whereas the constraints thesis is primarily concerned with the process of agricultural restructuring *within* which the family farm is transformed, the resilience thesis seeks to identify the 'logical features' of family farms as a distinctive *form* of production. Friedmann's contribution centres on the construction of a unitary concept of simple commodity production (SCP), characterised by:

(i) Family ownership and control of the agricultural means of production.
(ii) A family labour process with no, or strictly limited, use of wage labour.

(iii)   Family control over the means of reproduction, that is over the reproduction of labour power.

The inherent stability of this form of production is central to her concept of SCP and her explanation for the survival of the family farm. This stability, she argues, derives from the

> integrity of the household as a unit of production and consumption [which] means that there is no structural basis for a division of the product . . . and therefore no structural requirement for a surplus product (i.e. profit). (1981, p. 14)

As a result, the family farm is insulated from the law of value, and from the capitalist imperatives of expanded reproduction and capital accumulation. Instead, 'under intensive competition, it is possible, if necessary, to work harder and consume less in order to survive' (Long, 1984, p. 4).

Friedmann has constructed her concept of SCP with impressive theoretical rigour. Hers is unusual amongst other attempts to theorise the distinctiveness of household or domestic production systems, in that she locates her concept of SCP firmly *within* the capitalist mode of production (1981, pp. 16–17), rather than in any sense as 'articulating with it'. This examination of her work focuses on her concept of reproduction as central to the analysis of the dynamics of the family farm, and of the agricultural restructuring process more widely. The family farm, she argues, is more than a unit of agricultural production. It is a social unit of consumption with an internal dynamic in the process of daily and generational reproduction. She identifies the dependence of the agricultural production process upon the reproduction of labour power and of the relations of production. The reproduction process thus conceived is located within the family household's capacity to replace and supplement family labour from within its own ranks and to reduce the costs of reproducing labour by combining productive and personal consumption (1981, p. 16).

There are three aspects to her analysis on this point that are of particular value. She gives analytical significance to the family household component of the 'double specification' of the internal relations of the family farm referred to earlier. In particular, she distinguishes two sets of interdependent relations between the family household and the simple commodity form of agricultural production. First, the reproduction of family labour, on a daily and generational basis, and second, the reproduction of the relations of production of the SCP enterprise through the transfer of capital and land between successive generations of the family (1980, 1986a). It is these features which Friedman argues constitute the basis for the flexibility of the SCP enterprise and its competitive advantage over capitalist producers (1978a, 1981). They stand in contrast to the relations of reproduction of labour and the social relations of production in a capitalist enterprise where these are commoditised and dependent upon the wages and profits determined in the process of competition and, therefore, not internally regulated (1981, 1986b).

In specifying more precisely the internal dynamics of the family enterprise, Friedmann's analysis is an important step forward in theorising the distinctiveness of the family farm as a 'unity of capital and labour' (1986b). However her analysis faces a number of serious problems. Critiques of Friedmann's work have focused on the universality of her concept of SCP which is of limited analytical value in understanding the diversity and transformation of specific historical forms of family farming in advanced capitalist societies (Goodman and Redclift, 1985; Bernstein, 1986). As has already been argued,

> in imposing an internal consistency on SCP in order to buttress its use as a conceptual and empirical category ... the danger is of conferring on SCP the status of a theoretical category whereas it is an historically contingent phenomenon. (Goodman and Redclift, 1985, p. 23)

Moreover her analysis reinforces a rigid division between two monolithic categories, or forms, of agricultural production, family farms (SCP) and capitalist farms, permitting no differentiation within them. Part of the problem here derives from the level of abstraction at which the concept of SCP is couched and its relationship to a confusing array of other terms for types of household production.[7]

However her concept of reproduction is also highly problematic, in ways familiar in other Marxist debates concerned with the status of women's domestic labour (Himmelweit and Mohun, 1977; P. Smith, 1978). These are neatly summarised by the observation that

> an over-simplistic reading of reproduction leads to an assumption firstly, that social systems exist to maintain themselves through time [reproduce themselves] and secondly, that all levels of the system must be maintained through time in the same way. (Edholm *et al.*, 1977, p. 103)

In Friedmann's theory of SCP, reproduction is a non-transformative process, so that change over time or differentiation in the internal and external relations of the family farm are not permitted. The net result is a concept of reproduction akin to the functionalist concept of 'static equilibrium', which raises the question 'if the elements of [SCP] all gear into one another successfully, how and why does change occur?' (Bridenthal, 1979, p. 107).

There are two related aspects to this problem. First, in seeking to avoid the pitfalls of 'articulation theses' by posing her concept of SCP within the capitalist mode of production at the outset, Friedmann's treatment of the relations between the family farm and the wider agroindustrial complex is underdeveloped.[8] The family farm is viewed as a production unit located in, but insulated from, the imperatives of capitalist expansion. She theorises the commoditisation process in terms of the circulation of agricultural commodities in the market, rather than of the transformation of the agricultural relations of production in the restructuring process (Chevalier, 1983). As a result, she deals inadequately with the transformation of the technological relations of production, by which family farms are tied into wider circuits of agroindustrial and banking capital in such a way that the production process cannot be reproduced outside them.

A more satisfactory 'model' of reproduction is provided by van der Ploeg in so far as it situates reproduction within a conception of commoditisation as the transformation of the social relations of production. Figure 2.3 shows his model of the transformation of simple commodity production, in which the means of production are reproduced 'autonomously', 'through the agricultural production process itself', by increasing dependence on technological inputs which necessitates 'market dependent' reproduction (1986, pp. 37–8).

In effect, Friedmann equates the internal stability of SCP with the stability of its position in relation to the wider agricultural economy. However, without expanded production and the capital investment which this necessitates, in a competitive and technologically progressive production arena, such farms are destined to increasing marginalisation in terms of their economic contribution to total agricultural output and their share of total agricultural capacity (i.e. land area and fixed assets). While this does *not* mean that family farms are doomed to disappear (they may survive by reducing levels of household consumption or diversifying sources of money income), it does make Friedmann's argument that SCP enterprises are more

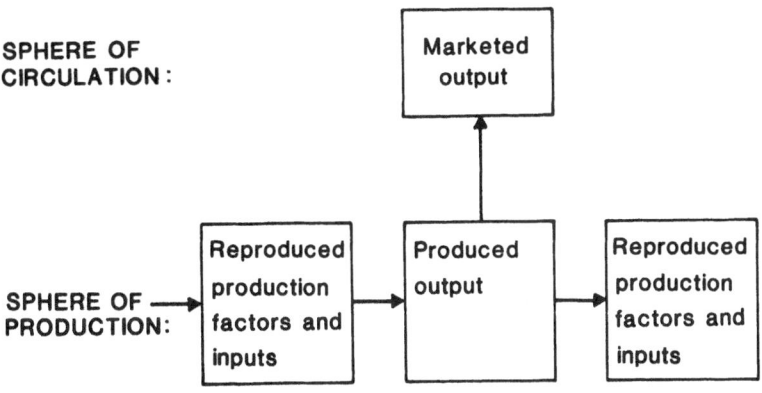

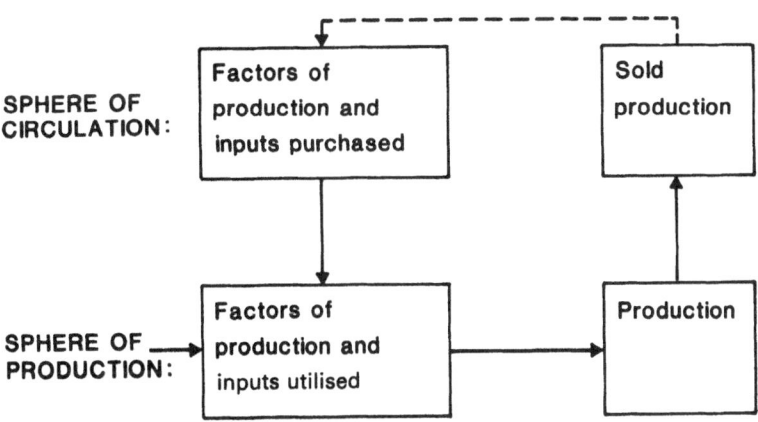

Source: Van der Ploeg (1986) pp. 37–8.

Figure 2.3   Autonomous and market-dependent reproduction

competitive than capitalist ones difficult to sustain (see also Reinhardt and Barlett, 1989).

This exposes a second problem with Friedmann's treatment of reproduction. Her analysis of internal stability relies on a notion of 'self-exploitation' *within* the SCP enterprise as the means by which this form of production is reproduced intact. She argues explicitly against earlier analyses which cast SCP as an exploited class. Reducing levels of household consumption has been clearly shown to be an important means of redirecting surplus to sustain the agricultural production process in competitive conditions (for example, Kahn, 1982; Marsden *et al.*, 1986b). However, as Bernstein has recently noted, the concept of *self*-exploitation to describe this strategy is embedded in 'residual, unproblematised and unitary notions of family and household' (1986, p. 18), a theme that will be explored further in the next chapter.

As with Goodman and Redclift, and despite her focus on the internal relations of the family farm, there are still no *people* inhabiting Friedmann's SCP enterprise and, crucially, no analysis of the social divisions and power relations which structure their position within it. These general problems are partially taken up in recent contributions trying to take the analysis of agrarian PCP beyond the terms of the 'constraints' and 'resilience' dialogue.

RECENT DEVELOPMENTS

Both the constraints and resilience thesis have been increasingly identified as inadequate as social theory. There are two related elements to this critique, familiar in wider 'post-structuralist' debates. The first is their inability to deal with empirical diversity. As Scott has argued:

> The reduction of historically specific situations to the basic algebra of value theory tended to produce oversimplified, essentialist models that belied the rich variety of concrete forms [of PCP]. (1986a, p. 3)

Attention has focused on the differential extent of commoditisation in agriculture, regionally and sectorally, and the consequent diversity of forms of family farming in terms of the manner and degree to which they are integrated into the wider market economy (Kahn, 1982; C. Smith, 1986; van der Ploeg, 1986). The second element is the inadequacies of such analyses in their treatment of human agency, ideology and lived experience. Long, for example, has argued that their fundamental shortcomings 'revolve essentially around the lack of attention given to the active role played by peasants, farmers and small-scale entrepreneurs in the process of commoditisation itself' (1986, p. 2).

Critics have highlighted the narrowly 'economistic' terms of analysis in which the social relations of the family farm have been reduced to a distinctive configuration of capital and labour. In the case of the constraints thesis this is manifested in the functionalist treatment of the family farm, which takes its internal relations as given. In the case of the resilience thesis, it is manifested in the reduction of the internal relations of the family farm to its features as an enterprise rather than as a family household. In both cases 'the family' and its constituent social relations are marginalised from the analysis. These two strands of criticism are interrelated in the sense that it is primarily through the mediation of human agency that the commoditisation process is seen to be modified and realised differentially between farms in particular times and places.[9]

The theme of differentiation amongst family farms in terms of their 'level of commoditisation' has recently been developed through the analysis of the diversity of production relations associated with family farming in the restructuring of contemporary English agriculture. In particular, a research team in which the author has been involved,[10] has been centrally concerned to integrate the analysis of the internal and external dimensions of farm production relations (Marsden *et al.*, 1986a). In opposition to Friedmann we have suggested that SCP is only one of a *range* of farm types based on family ownership of the means of production and family labour in the labour process, involving a variety of compromises with external capitals. In response to Goodman and Redclift we have argued that the subsumption process cannot be conceptualised as an undifferentiated, inexorable imposition of capitalist relations of production onto family farms, but must take account of the modification of the restructuring process by the active participation of individual farm families with their own 'non-economic' motivations.

The key to understanding this diversity and the process of farm business change is argued to lie in the interaction between the familial and enterprise dynamics of family farms and the extent to which they are tied into the wider capitalist economy through links with non-farm capitals involved in the agricultural production process. To this end, a realist methodology was used to develop a 'relational', as opposed to 'taxonomic', typology of farm businesses.[11] Four 'ideal types' of farm, three of them types of family farm, are defined as the core of a matrix of possible ways in which farms are directly or

indirectly commoditised. Direct commoditisation involves the transformation of the internal relations of production on the farm (capital
ownership, land-ownership, management organisation and labour
relations) in such a way that they move from a family-based to a
corporate-based form. Indirect commoditisation involves the development of external relations with fractions of capital elsewhere in
the agrofood complex (agro-inputs firms, food processing and
retailing firms, and credit institutions). (See Whatmore *et al.*, 1987a,
1987b, for a full exposition of this method.) The principles of this
typology are adapted in the methodology developed in Chapter 4 for
the analysis of different gender divisions in the family labour process
of farms characterised by different degrees of commoditisation.

Running in parallel with this work has been research centred at
Wageningen, where efforts have been focused on getting to grips
theoretically with the question of human agency (Long *et al.*, 1986).
In particular, attention has been drawn to the importance of incorporating into an analysis of the farm labour and commoditisation
processes the social and personal meanings of work generated by
participation in production. As Long has argued,

> if one wishes to achieve a deeper understanding of the social
> relations of production within specific economic units, then one
> should attempt to gauge the social estimation of the value of the
> labour in question as expressed by the individuals, groups or
> classes involved. [These] relations ... cannot adequately be
> comprehended if one concentrates solely upon the problem of
> subsumption within the logic of capitalism. (1984, p. 12)

He proposes (1984, p. 10) a revised framework for the sociological
analysis of structural change with three components:

> (a) a concern for the ways in which different social actors interpret
> and manage new elements in their life-worlds;
> (b) an analysis of how particular groups create space for themselves
> within changing structural conditions;
> (c) an attempt to show how these interactional and interpretive
> processes can influence and are themselves influenced by the
> broader structural context.

These arguments accord with a growing appreciation in Marxist
writing more widely of the need for concrete analyses of the
restructuring process to address the subjective experience of work
and the ideological processes through which the social relations of

production are maintained and transformed (Sayer, 1983; Thrift, 1987). Such an approach is founded upon a recognition that

> People at work . . . produce ideas about their social relations . . . along with their products . . . The ideas developed in production (the labour process) are a part of the practices that structure and are structured by the social relations of production. (Burawoy, 1979, p. 9)

These recent contributions mark important steps towards developing a 'post-structuralist' agrarian political economy. Three points in particular are carried forward from the contemporary literature examined here into the theoretical framework and analysis presented below:

1 The process of agricultural restructuring is seen to be distinctive as a result of the biological nature of the production process. It is conceived of in terms of the commoditisation of the relations of production and reproduction of farming, both in its position within the 'agroindustrial complex' and in the internal farm labour process.
2 Commoditisation is taken to be a social process mediated by, and realised through, the active participation of individuals involved in the farming (and in other sectors of the 'agroindustrial complex'). As such, one of the principal features of the commoditisation process is the differentiation of the conditions and forms of family farming.
3 The everyday work practices of family members in the farm labour process, and the ideologies of work mobilised, and in part produced, through these practices are taken to be a crucial element in producing, sustaining and mediating transformation in the labour process.

However agrarian political economy retains a number of outstanding problems in terms of the analysis of the persistence and transformation of family farming, which recent contributions have failed to consider. These problems revolve around an inadequate conceptualisation of 'the family', the farm labour process and, particularly, the nature of 'family labour'. These inadequacies betray weaknesses in the Marxist political economy framework which cannot be resolved by adopting less structuralist and less economistic terms of analysis alone. They can, however, be addressed by a feminist analysis, outlined in the next chapter.

3 A Feminist Reconstruction

THE ELUSIVE FAMILY

More than a decade ago Friedmann argued that the project of explaining the survival of the family farm and of analysing the processes and relations through which it is reproduced '. . . requires a sociology of the family as a productive organisation' (1978a, p. 576). But she concluded then that the 'conceptual tools' necessary to the task were not available. While agrarian political economy has since developed on a number of fronts, 'the family' has remained conceptually and theoretically elusive. Traditional Marxist analysis has generally paid scant attention to the family because it was seen to be marginalised from the production process as capitalism developed (McDonough and Harrison, 1978). From this position the family came to represent a vehicle for reproducing the capitalist relations of production (Kuhn and Wolpe, 1978).[1]

However, as Rapp has argued, this kind of approach is 'conceptually wedded to an acceptance of a distinction between the family itself and the larger world' (in Rapp *et al*; 1979, p. 57) in which the historically contingent separation of social production and the family/household under western capitalism has been incorporated into the basic concepts of social theory (Zaretsky, 1976). As Rapp goes on to point out, it is this dichotomous conception of the family and the wider social formation which 'creates the problem of its insertion into the world' (1979, p. 57).

More recently these reductionist terms of analysis have been extended to the internal relations of the family itself, principally through the Marxist-feminist inspired 'domestic labour debate' (Coulson *et al.*, 1975). Here the gender division of labour within the family is examined in terms of its services to capital as a cheap means of reproducing labour power.[2] However, as has been widely noted elsewhere, whether or not the alleged function of unpaid domestic household labour for capital is established, no mention of *gender* is necessary to this theory of the internal relations of the family. Most seriously, it 'throws no light on why domestic labour is exclusively female, but rather assumes that it is . . . yet from a feminist or family sociologist's standpoint, this is precisely what is to be explained' (C. Harris, 1983, p. 194).

28

Feminist critiques of mainstream and Marxist analyses have been particularly influential in exposing their chaotic concept of the family (Barrett, 1980) and narrow concept of 'labour' and the 'economy' (Bradby, 1977; Garnsey, 1981). Following Rapp's arguments, this chapter presents an attempt to theorise the internal relations of the family as a productive organisation. It argues that Marxist social theory, which has informed agrarian research thus far, presents major obstacles to such an exercise and draws on the insights of feminist theory and research as a way forward (Nicholson, 1987).

Feminism may now have established a 'new orthodoxy' on the critical social research agenda (Bowlby *et al.*, 1986, p. 321). However its application to the analysis of family farming, as a form of domestic commodity production in advanced capitalism, calls into question some of the orthodoxies of feminist theory itself. Informed largely by an urban research context located in the wage-labour economy of advanced capitalist countries, feminist theory does not provide a readily transferable theory of the family which can be slotted into agrarian political economy to meet Friedmann's prescription. However it does provide a conceptual basis for developing political economy analyses of the internal relations of 'the family' beyond the reductionist terms of orthodox Marxism. The central concept drawn from this perspective is that of *patriarchy* as a means of theorising gender relations in which 'women are inscribed in unequal, passive and subordinate power relations to men' (McRobbie, 1982, p. 48).

The framework set out below adapts arguments more familiar in the literature on household economics in developing countries than in the family farm context (Mies, 1982; Moore, 1988). It is guided by an important lesson drawn from that literature, namely that

> . . . while an analysis of the relationship between the family and the labour process must be central to any account of the mode of production as a whole, . . . the form taken by household labour, and the nature of its appropriation are historically specific . . . generalisation about a universal set of domestic relations, in which women share an ahistorical, acultural similarity, obscures more than it reveals. (Redclift, 1985, p. 94)

The first section below looks at the problems confronted by Marxist political economy in making sense of the internal relations of the family household as a commodity producing unit. The second introduces the concept of patriarchal gender relations and re-examines the social relations of domestic commodity production in the light of it. In the final section, key points taken from this re-examination are combined with those carried forward from Chapter 2 and used to construct a revised political economy framework for the analysis of the family farm. In the framework put forward, patri-archal gender relations and capitalist class relations are seen to intersect in specific, dynamic and complex ways, constituted through social structure and practice.

BEYOND THE UNITY OF CAPITAL AND LABOUR

Following Friedmann's recent autocritique of her earlier work (1986b, 1986c, 1986d), the terms of orthodox Marxist analysis can be seen to present two main obstacles to analysing the internal relations of the family farm. The first difficulty is rooted in the concept of petty or simple commodity production which lies at the heart of agrarian political economy. Marx's definition of SCP exposes these roots particularly clearly.

> The simple commodity producer is cut up into two persons. As owner of the means of production he is a capitalist, as labourer he is his own wage labour. As capital . . . he exploits himself as wage labourer, and pays himself in surplus value. (Marx, 1969, p. 408)

Simple commodity produc*tion* is here conceived of as an individual produc*er* – itself defined, significantly, in masculine terms. This linguistic sleight of hand effectively renders the composite nature of the social relations of the family/household analytically redundant. The crux of the problem for Marxist political economy lies in the fact that SCP 'does not include either families or households in its con-cept; it is simply the unity of property and labour within the context of generalised circulation of commodities' (Friedmann, 1986c, p. 48, my emphasis).

From this perspective the unsatisfactory nature of Friedmann's original concept and analysis of the reproduction process, examined in the last chapter, can be more clearly specified. The daily capacity to reproduce family labour and the generational capacity to reproduce family property are posed as important and distinctive features of SCP. Critically, however, they are not thought of as active processes, whether as part of the farm labour process in the production of subsistence goods or as the bearing and rearing of heirs. For example, Friedmann refers to the stability of the social relations of simple commodity production in terms of a '*self*-perpetuating' structure (my emphasis) (1978a, p. 12). Likewise, her analysis of the reproduction of labour power is couched in abstract terms, as 'the absence of a division of the surplus' (1981, p. 26). Even van der Ploeg's (1986) more sophisticated analysis casts reproduction in terms of the capacity of the structure to perpetuate itself. Simple commodity production no more 'perpetuates itself' than a simple commodity producer 'exploits himself'. What is lacking is a concept of 'the family' and a theory of its internal social relations which structure the processes of production and reproduction on the farm.

Friedmann's later work seeks to locate the family more centrally in her framework of analysis. However it is incorporated in an 'off-the-peg' fashion and, as Scott has argued, 'remains relatively untheorised in terms of kinship . . . there is an underlying assumption that the family is to be equated with the nuclear household' (1986a, p. 5). Nor does recent work proposing an alternative concept of domestic commodity production (DCP) entirely escape this criticism (Hedley, 1981; C. Smith, 1984), despite the family being central to it in a way it is not with SCP. Thus DCP is defined as

> . . . embedded in sets of non-commoditised relationships through family and community ties but . . . depend[ent] on articulation with commodity markets to realise the value of what is produced and to acquire both personal consumption goods and means of production. (Curtin, 1986, p. 75)

Both SCP and DCP conflate two dimensions of the family which need to be analytically distinguished – kinship and household – and reproduce elements of familial ideology by taking it as 'axiomatic that economic relations within the domestic group are based on pooling, sharing or distribution [and] defined in terms of a unified circuit of production and consumption into which exchange does not intrude' (O. Harris, 1982, p. 145). This is problematic in so far as it takes for granted what needs to be empirically investigated and incorporates a particular configuration of kinship/household relations as a universal general concept (Edholm, 1982).[3]

Friedmann's analysis of simple commodity production for a long time implied a division of labour within the family household. In an early paper she refers to the necessity for a 'harmonious, but not necessarily equal division of a family labour' (1981, p. 14). More recently, however, she has argued quite explicitly that 'the division of labour, patterns of domination and struggle, the cyclical life of the enterprise are all shaped by gender and generation' (1986a, p. 187). However her subsequent analysis tells us very little about *how* gender and generation 'shape' the farm labour process. Again, the same can be said of analyses utilising the concept of domestic commodity production. Despite their emphasis on the significance of 'non-commoditised' family relations, these are no more clearly theorised than in the case of SCP, except in the negative sense of not being commoditised. In both cases, where the analysis of the internal relations of the 'family farm' is developed any further, the focus has been the generational process of reproduction in terms of the transfer of capital through the social institutions of kinship, filiation and patrilineage.

The gendered basis of the family/household division of labour, while now admitted as an empirical fact, is nowhere integrated into the analysis of SCP or DCP at a *conceptual* level.[4] Thus, as a contemporary commentator has complained, 'despite [more] frequent mention of patriarchy, family labour and household relations . . . little [is] known about the power relations involved within the SCP labour process or the ideology that underlies them' (Scott, 1986a, p. 5).[5] The consequence of this failure to theorise familial gender relations has been that women's labour is lost from view because it '. . . fall[s] down the divide between agricultural work on the one hand and the relations of kinship and filiation on the other, between "production" and "reproduction"' (Mackintosh, 1979, p. 176).

Friedmann's attempts to add a gender dimension to her existing analysis brings to the fore the second, and more fundamental, aspect of the difficulties facing Marxist political economy. Like family and household, the social divisions and relations which constitute them fall outside the Marxist conceptual schema. What kind of social relations are gender and generation? Are they an equivalent kind of social division as Friedmann's analysis suggests? If not, what are the different practices and structures through which they are constituted? The obstacle to answering these questions, indeed to raising them, in orthodox Marxist analysis is laid bare in Friedmann's autocritique, but remains unresolved in terms of her subsequent analysis. She argues that

it is difficult to analyse family and SCP in conjunction with one another because generalised commodity circulation individualises human beings. It shapes most people as bearers of labour power,

. . . it shapes others as property owners. Therefore the unity of property and labour refers logically to individuals. (1986c, p. 50)

There are two, related components to the difficulty posed. The first is that the orthodox Marxist concepts used to analyse capitalist social relations shape individuals only as labour or capital. Other social divisions such as 'race' or gender are not encompassed at a conceptual level. The second is that, as Friedmann points out, 'household and family are particularistic relations . . . individuals cannot be substituted one for another as easily as can employees or clients' (1986c, p. 51). In other words, where commoditisation individualises social relations, household and family particularises them in such a way that individuals are structured in their relationship to one another and to the social group as a whole in non-equivalent ways. As a result, not only are the social persons of husband and wife not interchangeable (i.e. across the gender divide) but neither are, for example, those of wife and daughter (i.e. within the same gender category).

Precisely because the concepts necessary to deal with theorising gender relations lie outside orthodox Marxist analysis, Friedmann's solution to the problems which she exposes represents a retreat rather than an advance in theorising the internal relations of the family farm. In order to preserve the integrity of the concept of simple commodity production she suggests that the solution lies in separating the analysis of the enterprise and the family, reserving SCP for the analysis of the enterprise and turning to the concept of 'survival strategies' to examine the family (1986d, p. 125). This undermines the whole basis of her earlier (and more general) arguments about the need to comprehend the social relations of 'the family as a productive unit' as the crux of explaining the persistence of family farming under capitalism. But the alternative is to admit that SCP, as the *unity* of capital and labour, falls apart under closer scrutiny because of the highly *disunited* nature of family labour relations, constructed through unequal gender divisions in the labour process and in the property rights controlling the means of production. The concept of domestic commodity production is less fundamentally flawed in this respect, as it accommodates the unity of household and enterprise, but as yet the internal social relations which structure this unity remain untheorised (but see Brass, 1986).

A theoretical framework is required which 'problematises' the concepts of production and labour employed in analysing the family labour process and which 'deconstructs' the family household unit in order to analyse the principal social division structuring its productive and reproductive relations – gender. It is the lack of a theory of gender relations which underlies the difficulties confronted by Marxist political economy in analysing the family farm. Gender relations cannot be understood in terms of Marxist value theory, when, as Bradby has pointed out, 'the labour of half the world does not take the value form' (1982, p. 125), nor in terms of Marxist class theory, when, as Hartmann has argued, the social categories of capital and labour are 'sex-blind' (1979, p. 8).

This is not to argue either that gender relations can be understood outside, or apart from, the specifically capitalist structure of the wider social formation, or, like Delphy (1984), that it is patriarchy rather than capitalism which represents the dominant structuring mechanism of that wider social formation. Rather, it is to suggest that gender relations are not derivative of, nor reducible to, capitalism and that the burden of explanation needs to be shifted to an *intrinsic* theory of gender relations which comprehends the power relations between men and women *qua* men and women, and the practice and experience of women's subordination to men within the family household as a production unit.

PATRIARCHAL GENDER RELATIONS

In focusing on the concept and analysis of gender as a social division, feminist theory has sought to challenge the notion that this division is natural. The argument has not been that biological differences between male and female are inconsequential but that

> it is not sexual differences as such which explain social relations between men and. women, but rather the social usage and social meaning which is attributed to them and legitimised by their attribution as 'natural facts'. (Stolke, 1981, pp. 174–5)

Rather than simply elaborating or expressing 'natural' differences,[6] social processes, through a range of institutions, codes and taboos, mediate and modify biological differences and inscribe them in the gender identities of masculinity and femininity (Rubin, 1975; Connell, 1985). This distinction between sex and gender is a critical one in the development of explanations of the relations between men and women. Feminist work has shifted the terms of debate away from biological determinism (Bartells, 1980) and from atomistic social theories, in which gender is treated as an individual attribute (Andre, 1985), towards a conception of gender as a socially constructed axis of human relations.

The transition through theories of 'sex roles' and 'gender roles', as natural or socially inculcated behaviour patterns, to 'gender relations', as power relations between men and women, has been well documented (Ayim and Houston, 1985; Connell, 1983). Role theory is untenable as social theory because it fails to comprehend the empirical diversity of gender roles over time and place and because, as a voluntaristic theory of social relations, it is unable to theorise power and social interest within which the social meaning attached to gender roles is forged. Research into gender relations focuses instead on the social institutions and structures through which unequal power relations between men and women are created and sustained. Walby (1986) has usefully categorised feminist theories of gender inequality, involving different interpretations of the concept of patriarchy, into three groups (p. 5).

– gender inequality as a result of an autonomous system of patriarchy, which is the primary form of social inequality (see, for example, Coward, 1983; Delphy, 1984).

– gender inequality as a result of patriarchal relations so intertwined with capitalist relations that they form one system of capitalist patriarchy (see, for example, Vogel, 1983; McDowell, 1986).

– gender inequality as the consequence of the interaction of autonomous systems of patriarchy and capitalism (see, for example, Mitchell, 1971; Hartmann, 1979).

From the arguments in the last section it will be clear that the position adopted here falls broadly into the third of these categories, but not without a number of important qualifications. Much work on patriarchy suffers from the structuralist tendencies criticised earlier in Marxist political economy in constructing a general theory of patriarchy as a logical system rather than as an active social process. As Foord and Gregson have argued, such theories are misdirected, because they imply that 'gender relations are . . . particular expressions of patriarchy [when it is] patriarchy that is a particular form of gender relations' (1986, p. 193). Moreover, as Connell (1987) has pointed out, the very terminology in which general theories of patriarchy are couched betrays a conceptual slippage into what he calls 'categoricalism' (p. 54). Biological sex differences, and specifically the fact of 'maleness', get attached *by definition* in such theories to the social fact of power. Theories of patriarchy, as a universal system of *male power*, therefore reinforce, rather than challenge, a biological explanation of gender inequality (Eisenstein, 1982).[7]

A conception of patriarchy as an autonomous social structure in which women are subordinated to men thus requires closer specification. At what level of abstraction is autonomy located and with what consequences for the analysis of the relationship between patriarchal gender relations and other social structures, such as capitalist class relations, at lower levels of abstraction? A rigorous attempt to tackle these difficult questions is made by Foord and Gregson who, working within a realist mode of analysis, propose a 'hierarchy of abstraction' (1986, p. 198) comprising three interrelated levels of analysis: general, particular (historically contingent), and individual or personal (Gregson and Foord, 1987, p. 373). This framework is not without its problems,[8] but it does provide a coherent basis for a non-reductionist analysis of patriarchal gender relations. Following their elaboration of this framework it is at the general conceptual level that gender relations and mode of production can be claimed to have autonomy as distinct objects of analysis. In consequence, patriarchy and capitalism, as particular historical forms of these general objects, are *conceptually* independent and contingently rather than necessarily related at the level of historical analysis. It is at the level of concrete analysis that patriarchal gender relations and capitalist class relations mediate one another through the daily lives of individuals in specific circumstances in such a way that they are manifested interdependently in a multiplicity of ways (Foord and Gregson, 1986, p. 201).

Foord and Gregson claim that the interrelationships between these three levels of analysis 'encapsulate [the] movement from the most basic characteristics of an object through to the specific and various ways in which these properties are manifested in lived experience' (1986, p. 198). However their approach provides for little reciprocity in the relationship between them. In particular, it allows little room for the active role of social practice and individual consciousness at the concrete level of analysis to inform the more abstract levels of a theory of patriarchy.[9]

In the light of the arguments presented at the end of the last chapter this framework is modified here, drawing on a set of analytical concepts proposed by Connell which enable patriarchal gender relations to be examined as a *transformational* process 'generated by the interplay of social practice and social structure' (1987, p. 53). He distinguishes between *gender order*, as the macro-level historically structured power relations between men and women, and *gender regime* as the micro-level stage of gender politics played out through the practices of particular households, workplaces and so forth (1987, p. 111). The relationship he proposes between these two levels is a dialectical one in the sense that practices within particular regimes may depart from or even contradict the macro-order of gender relations. Such contradictions may provoke 'policing' (Donzelot, 1979) or they may signify structural tensions which could lead to the transformation of gender relations through concerted social action (Connell, p. 114). Within this framework, the relations and practices of the gender regime of 'the family' can be examined as an integral part of a wider system of patriarchal gender relations, rather than as an isolated unit of analysis.

In common with much feminist research, both Foord and Gregson, and Connell identify the principal site of gender relations (and hence patriarchy) as the process of human reproduction. Foord and Gregson locate the material basis of patriarchal gender relations in 'biological reproduction and heterosexuality' (1986, p. 202), and men's control of women's fertility and sexuality, while Connell specifies the processes of 'engendering, childbirth and parenting' (1987, p. 140). However, although reproduction, like production, may be regarded as a universal, it is always embodied in culturally and historically specific forms (C. Harris, 1983, p. 182).

Yet feminist theories of patriarchal gender relations are based on highly universalistic concepts of the family as the social institution central to the organisation of the process of human reproduction. This is problematic, particularly for the analysis of the family as a productive unit. One common problem is the elevation of the nuclear family household to the status of a universal standard against which all domestic groups are interpreted.[10] A second problem, more deeply rooted and implicit in both Foord and Gregson's and Connell's analyses, is the conceptual separation of the processes of reproduction and production which non-western experience and PCP confound (Bland *et al.*, 1978; Beneria, 1979). The next section attempts to revise the concepts of reproduction and 'the family' in such a way that they are appropriate to theorising gender relations in the context of domestic commodity forms of production in advanced capitalist societies.

Reproduction as labour process

The historical separation of subsistence production from social production with the emergence of generalised commodity exchange and wage-labour, characteristic of advanced capitalist countries, has been widely translated into a conceptual separation between the mode of production (or economic level) and associated class conflict, and other levels which function to reproduce that relation (Edholm *et al.*, 1977). This dualistic conception is outlined in Figure 3.1.

LEVEL OF ABSTRACTION	PRODUCTION	REPRODUCTION
GENERAL	MODE OF PRODUCTION	MODE OF REPRODUCTION
NECESSARY RELATIONS	LABOUR MEANS OF PRODUCTION	BIOLOGICAL REPRODUCTION HETEROSEXUALITY
HISTORICAL FORM	CAPITALIST CLASS RELATIONS	PATRIARCHAL GENDER RELATIONS
CONCRETE SITES	ECONOMY/EMPLOYMENT	FAMILY/MARRIAGE

Figure 3.1   Dualistic approaches to production and reproduction

Such a framework obscures the essential interdependence of these two processes and loses sight analytically of the household livelihood practices which bind them together.[11] Where the Marxist preoccupation with capital accumulation or expanded reproduction has overshadowed issues of human reproduction (Bennholt-Thompson, 1982), much feminist analysis reduces human reproduction to a biological process. The generational process of procreation (biological reproduction) is disassociated from that of acquiring daily sustenance (Redclift, 1985, p. 118) and from the productive labour which this process involves. Drawing on the work of Edholm *et al.*, (1977) and Bennholt-Thompson (1982), Figure 3.2 outlines the component processes underlying the composite concept of reproduction. The relationship between these processes and the social relations which structure them are historically variable.

Following from this more complex concept of reproduction, the productive labour process can be seen to span both the processes of production for subsistence and for 'social exchange'.[12] Moreover the

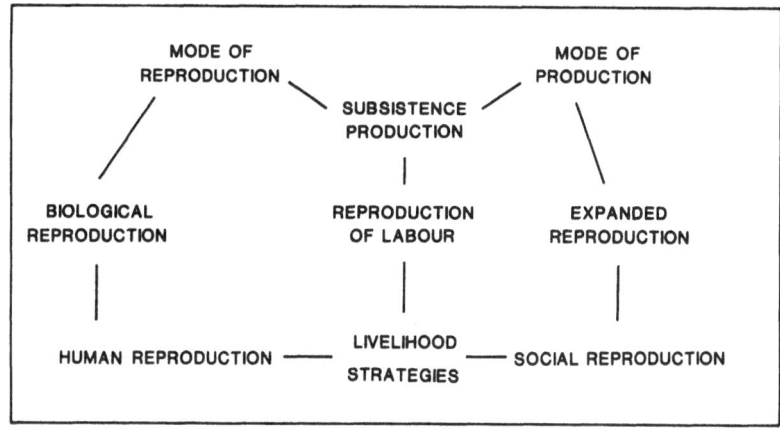

Figure 3.2   Production, reproduction and livelihood

social organisation of labour within and between these processes constitutes another important site for the construction of patriarchal gender relations through inequalities in terms of access to the means of production/subsistence, the division of labour, and through the appropriation of women's labour and its products by men (Hartmann, 1979; Delphy and Leonard, 1986). Work in the subsistence circuit of production has been variously termed 'reproductive' or 'domestic' labour, lumping together a range of tasks, from childcare to food preparation, which sustain the daily and generational reproduction process. However such terminology universalises the particular conditions of advanced capitalist societies, where subsistence production is mediated by a wage, and distorts the analysis of work performed by women across a range of production systems where the wage-labour economy does not pertain (Bradby, 1975; Beneria, 1982).[13] The family farm represents one such case, where domestic labour, as non-commoditised labour, is characteristic of all household members' labour, across agricultural as well as subsistence production, and is not restricted to women's 'domestic' work.

A second point to follow from this revised concept of reproduction is that reproduction and production form a unity within which some degree of autonomy needs to be accorded to the relations of reproduction (Redclift, 1985, p. 119). Such an approach breaks out of dualistic conceptions of production and reproduction as separate 'spheres' of activity, treating them instead as interlocking processes which can be seen to be structured simultaneously by gender and class relations. Most importantly, in terms of this research, it liberates 'the family' from a narrowly defined reproductive role. Instead, the particular gender regime of the family, as an historically and culturally variable social institution, can be examined within the context of a wider gender order constructed through a number of practices including sexuality, gender identity, physical and emotional power, and divisions of labour and property, sited in a range of social institutions. The family itself can now be examined in more detail as a social institution fundamentally structured by gender and which, in turn, is a key site for the construction and contestation of patriarchal gender relations.

The conjugal household

The starting-point for feminist analyses of the family is the fact that

> *the* family does not exist other than as an ideological construct, since the material relations of which it is composed – household, kinship and familial ideology itself, are historically, socially [and culturally] specific and diverse. (Barrett, 1980, p. 199)

Thus to use the term 'the family' without qualification is to collude in the reproduction of a familial ideology which reifies the nuclear family as a natural, timeless unit of social organisation (Flax, 1982; Bernades, 1985).[14] Feminist anthropological work in particular has sought to specify more clearly the material relations and ideological practices which constitute the family in particular historical circumstances (Rapp, 1982). Returning to earlier arguments, two analytically separate components of the family – kinship and household – need to be distinguished. While neither of these concepts is unproblematic, they provide a more coherent basis for the analysis of the gender relations and practices associated with 'the family' and for understanding the ways in which specifically patriarchal forms of these relations are structured, contested and reproduced.

Kinship refers to two principles of social organisation, ties of blood (filial relations) and ties of marriage (affinal relations). However, as Connell (1987, p. 79) points out, 'kinship is not a list of biological relatives but is a system of categories and statuses which often contradict actual generic relations'. Kinship relations are the main structural system in control of the processes of bearing and raising children and a key site in the construction of gender identities. The vocabulary of kinship is used to invoke obligations, to determine rights of inheritance and in effect to legitimise the structure of the relations of production and distribution of power and authority within the kin group (Hedley, 1981, p. 76). Thus the crucial issue is to understand the ways in which marriage and filiation lend support to, and serve to perpetuate, social inequality and relations of power (C. Harris, 1983, p. 66).

Feminist research has established gender as a primary axis of social inequality and power relations within kinship structures (Yanagisako, 1979). Attention has focused on marriage as a social institution by which men exercise control over women's fertility and sexuality within historically specific procreative practices (Comer, 1982; Barrett and McIntosh, 1982), and as a key element in the structure and transfer of property between and within kin groups and hence in producing and reproducing social class (Hirschon, 1984). As Whitehead (1984) has pointed out the relationship between ideas of property and ideas of the individual person are intertwined and kinship systems play a key role in constructing men and women differently with respect to this relationship in ways which inhibit women's status as 'full persons'. Thus, where traditional and Marxist analyses have focused on filial relations and, particularly, the generational transfer of property, feminist research has pointed out that these relations too are fundamentally gendered and play a key role in reproducing gender inequalities over time.[15]

Household refers to the socioeconomic unit organising the subsistence process and centred on co-residence and commensal resource provision and consumption. It is a problematic concept on two counts: first, in defining the boundaries of the unit when commensality may vary between different resources and may not be fixed (C. Harris, 1981); second, and perhaps more importantly, its internal relations are frequently assumed to be non-economic, so that exchange, let alone the possibility of unequal exchange between household members, is not raised (Whitehead, 1981). Feminist research has stressed the need not only to be more circumspect in defining households, but for analyses to make explicit the power relations and social divisions which structure household livelihood in terms of the labour relations, divisions of goods and ownership of the means of production and subsistence (Mackintosh, 1981; Collins, 1985).

The relationship between household and kinship is very variable.

Thus, while household membership and the organisation of the subsistence process is commonly organised through kinship relations, the household is rarely coterminous with kinship, but is located within wider kin and community networks (Stivens, 1981). Moreover the relationship between household and kinship is likely to change over the life-course of any particular household in relation to major reproductive events such as marriage, births and deaths (Allatt *et al.*, 1987).

The term *conjugal household* is used here to define a particular combination of these sets of relations characteristic of family farms in advanced capitalist countries.[16] The conjugal household unit centres on a monogamous, heterosexual couple and is, by definition, structured primarily by gender relations. It is wider in compass than the more traditional concept of nuclear family household, comprehending a variety of possible 'phases' in the life-course of the couple at its core. Marsden (1979, p. 118) distinguishes five such phases: marriage and the setting up of a conjugal household; an expansion phase, associated with the birth of children; a dispersion phase, when children leave home; an independent phase, when the conjugal couple live alone following the departure of children; and a replacement phase when the farm is taken over by the children and the older couple retire off the farm. Within this framework it is the specific form of patriarchal gender relations, between men and women in the socially constructed roles of husband and wife, which becomes the focus of analysis.

The examination of feminist theory and research on gender relations here has by no means been exhaustive. Its contribution to constructing a framework of analysis for the internal relations of the family in agrarian political economy can be summarised under three points.

1   Patriarchy is a particular form of gender relations which represents a conceptually autonomous social structure, sited in, but not restricted to the process of human reproduction. Patriarchal gender relations constitute an active social process by which women are subordinated to men through a range of social practices and institutions.

2   Reproduction is taken to be a multi-layered process which is intimately related to production through the subsistence process. It involves productive labour which is characteristically gendered, but in which women's labour is carried out under historically and class-specific conditions.

3 'The family' is an ideologically-loaded composite of kinship and household relations which is historically and culturally varied in form. The concept of conjugal household, centred on a monogamous heterosexual couple but associated with life-course variations, is taken to define the intersection of kinship and household relations characterising family farms.

DOMESTIC POLITICAL ECONOMY

The foregoing sections suggest that moves towards 'post-structuralist' modes of political economy analysis examined at the end of Chapter 2 need to be reorientated. On the one hand, a focus on empirical diversity needs to incorporate not just variations in the degree of commoditisation, but also of household and kinship relations and the familial gender divisions within them. On the other, the exhortation to address the significance of human agency and the ideological dimension of the farm labour process needs to comprehend the familial gender ideologies within which men's and women's labour roles on the farm are constructed and legitimised. The framework suggested here thus recasts Burawoy's argument that 'the ideas developed in production (the labour process) are a part of the practices that structure and are structured by the social relations of production' (1979, p. 9) in terms of a wider conception of the economic. In particular, it focuses attention on the ideological processes and work practices through which women's labour is devalued in agricultural commodity production[17] and rendered invisible altogether in the case of household provisioning.

The term *domestic political economy* is adopted to describe the unity of the processes of production and reproduction, outlined in Figure 3.2. It captures the interdependence of family and enterprise, organised around the social relations of kinship and household, which define domestic commodity production. Drawing together the key points identified in the conclusion to Chapter 2 and the section above, these social relations can be specified more precisely as the intersection of patriarchal gender relations and capitalist class relations in the process of commoditisation. The basis of patriarchal gender relations has been identified at a theoretical level with the kinship ties of marriage, the household ties of economic dependence and the familial gender ideologies within which women are fashioned as 'wives'. However how these relations intersect with the process of commoditisation through family labour divisions and practices on the ground is a matter for empirical investigation.

The necessary components of an analysis of the domestic political economy of the family farm are thus: the household structure of those living and working on the farm; the composition of the principal and subsidiary labour divisions on the farm in terms of the participation of family and hired labour; linkages with external capitals in the farm production and reproduction processes; the circulation of income and money capital between them and the structure of the ownership of the capital and land assets on the farm between resident and non-resident kin (and non-kin members where appropriate).

In the light of the discussion above, the farm labour process needs to be reconceptualised to cover subsistence production, for household and family reproduction and consumption, as well as agricultural (and other) commodity production processes. However, as Pahl has noted in another context, '. . . various forms of work are inextricably intertwined, and analytic disentanglement poses formidable difficulties' (1984, p. 129). The term *labour circuits* is adopted here to avoid the difficulties associated with the concept of labour spheres, which suggests some kind of physical or spatial separation of activities (Gamarnikow *et al.*, 1983; Murgatroyd, 1985).[18] The family farm labour process can be conceptualised as potentially comprising four component labour circuits: agricultural labour; domestic household labour; non-agricultural farm labour; and off-farm wage labour.[19] The first two circuits may generally be regarded as the principal ones characteristic of farming, and the latter two as secondary, although this in itself is currently undergoing change with the growth of part-time and hobby farming. While it is useful to distinguish these circuits analytically, they are interrelated in practice. Between them they encompass all the subsistence, commodity and income-generating work undertaken by family members which may contribute towards the household livelihood and the reproduction of the enterprise, as well as directly to the production of goods and services for the market.

The principal axis of the gender division of labour is that between agricultural labour and domestic household labour. Domestic household labour consists of five main kinds of activity, which service the family household, reproducing its labour capacity and social relations of production on a daily and generational basis; these are: childcare, food provision, housework, laundry and shopping. Agricultural labour consists of those tasks directly related to the cultivation or husbandry of plants and animals and can be divided into two principal types, manual and administrative. The non-agricultural farm labour circuit consists of those on-farm activities producing non-agricultural commodities, such as farm shops, bed-and-breakfast catering and so on. 'Off-farm wage labour' describes that labour circuit involving farm family members in generating money-income for the household/farm budget through off-farm paid employment. The labour relations characterising each circuit require concrete analysis.

This framework provides a basis for analysing the way in which the gender regime of the family farm structures its activities as a unit of domestic commodity production, and the way this regime is exploited and reshaped by capital in the process of commoditisation.

4 Theory into Practice

Feminist research inevitably has repercussions for the kinds of methodological principles and techniques required to implement an understanding of social relations as always and everywhere gendered, and of gender relations as an active social process.[1] This chapter sets out the methods used to translate the earlier theoretical arguments into a workable means of collecting and interpreting information about particular family farms and farm women. In the first section, arguments for a methodology which takes account of women's subjective experience as a necessary part of understanding how patriarchal gender relations structure society are examined, through the case of family farming. A two-tier methodology, presented in the second section, is adopted in response to these arguments. The third section outlines the fieldwork which provides the basis of the analysis in subsequent chapters. It introduces the study areas, the farms and the farm women at the centre of the narrative.

MAKING WOMEN'S WORK COUNT

In the previous chapter it was argued that the treatment of labour in political economy analyses tends to be 'productionist'. This extends to the methods of evaluating labour which, constructed from the experience and structure of commoditised labour relations under capitalism, rely on measures of time and money (Thompson, 1967; Burawoy, 1979). When the concept of labour is revised to encompass non-wage forms these measures come to have little or no meaning (Beneria, 1982; Murgatroyd, 1985). Almost by definition, such work is unpaid and, as numerous studies of housework have shown, it does not conform to the regulated time regime associated with the capitalist working day, nor is it identifiable with a discrete workplace (Oakley, 1974; Berk, 1980).

Thus, even if the labour process is analysed within a theoretical perspective which renders such work visible, the problem of evaluating it remains (Beneria, 1982; Dixon–Mueller, 1985). Whatever the system of measurement adopted, the process of evaluation incorporates ideologies through which different kinds of labour (and labourer) are valued differentially. As Long has argued in a broader context,

46

'the analysis of non-wage labour ... inevitably raises questions of the social value and expectations of such work' (1984, p. 16). Any attempt to make women's work, as systematically devalued labour, count requires a methodology that elucidates the ideologies which legitimise exploitative labour relations and inform the experiences and *meanings* of work for women themselves (Roberts, 1981; Equal Opportunities Commission, 1986).

This raises a second and more fundamental methodological problem, the question of how to elicit and interpret the accounts of research subjects, as gendered individuals, whose everyday activities and ways of making sense of the world are structured within and reproduce the ideological apparatus of a patriarchal sex–gender order. These accounts give meaning to the world in different ways which, as discourses, embody prevalent power relations in such a way that the experience of dominant social groups (such as men) is more readily expressed than is that of subordinate groups (such as women) (Ardener, 1978).[2] How is the complexity of this web of social meanings and discursive practices, within which researchers and research subjects are (differentially) located, to be dealt with?

Pursuing the question of labour, a footnote on the methodology used by Symes and Marsden in their research into the role of farmers' wives illuminates a number of dimensions of this problem:[3]

> In most cases it was the farmer who answered the questionnaire. Ideally, of course, separate responses from both the farmer and his wife should be recorded and compared to determine the presence of bias. There is no reason to suppose, however, that farmers would consistently under- or over-estimate their answers in a particular direction. On those occasions where wives were present at the interview, their answers accorded closely with those of their husbands. (1983, p. 240)

The 'ideal' is to get both men's and women's separate responses in order to detect bias (see also Berk and Shih, 1980). The next best approach, which is the one Symes and Marsden adopt, is to ask the man, who it is supposed can have no reason to misrepresent his wife's work. Asking the woman herself, independently of her husband's verification, is not raised as an option. This is not to suggest that women are more likely to tell the 'truth' about their labour contribution or even, necessarily, to be more accurate about the 'facts'. Nor is it simply to make a point about giving women authority over the representation of their lives. Rather, in the light of the discussion above, it is to suggest that the limits of objective knowledge need to be recognised (Bourdieu, 1977) and account given to competing social realities, exposing the social interests on behalf of which they work (Strathern, 1984; Weedon, 1987). Such questions pose major problems for traditional positivist methods because they are predicated on an objective and stable social reality, and emphasise consensual rather than divergent experiences and values of that reality (Burgess *et al.*, 1988).

Clearly the questions raised here extend beyond the examples used to illustrate them and are situated within wider debates about research methodology.[4] The feminist contribution to this critical arena is refreshing in that it starts from a primary interest in challenging a specific, tangible outcome of the assumptions and techniques of positivist social research – the distortion of our understanding of women's lives – rather than being located at the abstract level in which the debate is more usually conducted. It has sought a methodology which validates both empirical objects and relations and their subjective meaning for individual actors as a means of allowing women's experience and self-knowledge to inform theory (McDowell and Bowlby, 1983).

In this context, feminist research has widely adopted qualitative or 'humanistic' approaches to collecting information. These include adapting survey techniques to increase their sensitivity to the discourse of research subjects; challenging the hierarchical relationship between researcher and 'respondent'; and according non-quantitative forms of information validity as 'data'. In terms of the problems highlighted above, qualitative methods offer some clear advantages over quantitative methods. But equally, as Graham (1983) has already noted, their ready adoption in feminist research also contains a real danger of methodologically 'ghettoising' women by implying that it is women as research subjects who require different, 'soft' data collection techniques. Moreover such methods themselves frequently embody positivist assumptions (Centre for Contemporary Cultural Studies, 1978) but, instead of reducing social reality to equivalent objective units, present it as '. . . reducible to and explainable in terms of the subjective meanings produced and deployed by actors in the concrete situations of face to face interaction' (Kuhn and Wolpe, 1978, p. 5).

Such approaches elevate what people say (a subject's discourse) to the level of pure data which is somehow contaminated by interpretative analysis. The problem with such material, as Geertz describes it, is that imprisoned in the immediacy of its own detail it is presented as self-validating, or, worse, as validated by the supposedly developed sensitivities of the person who presents it ... resisting conceptual articulation and thus ... escaping systematic modes of assessment. (1973, p. 24)

McRobbie identifies two particular misconceptions prevalent in such methods. Firstly, there is the notion that the subject's spoken word is in some sense *pure*, and that the faithful transcription of these utterances constitutes untainted, raw data. As she points out, there is no such thing as 'pure speech' (1982, p. 51). All speech is a product of a particular social context or set of social relations. The feelings and views expressed, and ways of expressing them, are peculiar to that context (see also S. J. Smith, 1981). Secondly, the act of recording speech as text (Hall *et al.*, 1980), however faithful the transcription, transforms it into an object or artefact which fixes things that in themselves are transitory, and thereby transforms their significance. The implication of these criticisms is that social analysis is unavoidably a *re*presentation of other people's representations of the social world.

MACRO AND MICRO ANALYSIS

The above discussion suggests that a methodology is required which enables the researcher to make sense of the everyday world of the research subject, within an analytic and interpretative framework which avoids the pitfalls of voluntarism, on the one hand, and of structuralism, on the other. This demands more than a revision of the techniques employed to elicit information which in themselves have both strengths and weaknesses. Instructive here are recent moves away from the ultimately unhelpful antagonism between structuralism and voluntarism towards a position, generically described as 'methodological situationalism' (Knorr-Cetina, 1981), associated with various attempts to develop a social 'theory of practice'.[5]

In different ways these attempts focus on the process of mediation between social structures and processes, and individual actions through a concept of *social practice*, describing people's everyday actions and experiences and the meanings mobilised and reshaped by individuals through them. Despite their differences, they share three principles for social theory which inform the approach adopted here.

First, that individual action is necessarily situated in wider social relations and can only be understood as 'interactionally accomplished' (Knorr-Cetina, 1981). Second, that social structure cannot be conceptualised as an entity independent from human activity or from the meanings attached to such activity by individual actors. Rather, as Sayer has put it '. . . while certain actions or ways of acting are conditioned by particular social structures, the existence and reproduction of those structures are contingent upon the execution of those actions' (1984, p. 32). Third, that social action is transformative and social structure can become an object of social action in itself, whether through class conflict or struggles against sexism and racism (Connell, 1987, p. 95).

An important distinction is made in such approaches between *social relations* as structural relations at the macro level of analysis and *personal relations* as the lived relationship between particular individuals at the micro level (Burawoy, 1979, p. 220). In attempting to inform the analysis of the family labour process by the everyday experience and meaning of work for women as farm wives, this distinction is central. Heller's concept of the process of 'everyday making sense' provides a theoretical route into the personal, representing it as a kind of second nature in which people orientate themselves without deliberate reflection so that lived experience presents what is socially produced as natural and beyond human control (P. Wright, 1985, p. 9). Thus she argues that, 'while everyday life is moulded and delimited by social structure, it does not in itself simply express that structure' (Heller, 1984, p. 218).

The two-tier research methodology adopted here is designed to explore both macro and micro dimensions of analysis. It consists of two complementary levels of data collection and interpretation – an *extensive*, descriptive level using survey methods and an *intensive*, explanatory level using ethnographic methods.[6] The purpose and contents of these two levels can be summarised as follows:

1   Extensive research: this level of the analysis adopts a survey data collection technique and attempts to provide a *descriptive* account of the pattern of familial gender divisions in the farm labour process at an aggregate level. In focusing on women's labour, it takes women's own assessments as the basis of analysis, adopting a system of measurement which counts all labour as productive, whether it produces use-values for household consumption or exchange-values for commodity markets. In this sense its empirical contribution is to

extend analysis of the farm labour process beyond the realm of agricultural commodity production and to describe the main features of gender inequality in the family labour process, in terms of the division of tasks and the labour relations under which women work. 2 Intensive research: this element of the analysis looks in greater depth at a limited number of case studies and attempts an *explanatory* analysis on two fronts. First, using ethnographic techniques, it seeks to provide an interpretive understanding of women's experiences as wives in the farm labour process, and an analysis of the different and unstable familial gender ideologies within which these experiences are understood. Second, informed by the analysis of labour ideologies and practices, the domestic political economy of six family farms is reconstructed, focusing on the interrelations between household and enterprise and the intersection of class and gender relations in the farm labour process.

FIELDWORK AND METHODS

A variety of methods of data collection and analysis were used to inform the two-tier framework outlined above. These are described in this section. Further, technical details, can be found in the appendix.

Study areas

The research was carried out in two areas in southern England, one in west Dorset and the other in the Metropolitan Green Belt around London (MGB). The geographical location of these study areas is shown in Map 4.1.

These two areas were selected because they represent very different kinds of farming systems and rural environments. The MGB is an area of mixed farm enterprises and varied scales of production. While agriculture is a conforming land use within green belts, the area is characterised by strong pressures on land and land prices from non-agricultural interests (Munton, 1983). These pressures arise in particular from residential and commercial property developers, but also from public developments such as the M25 motorway. Opportunities for non-agricultural and off-farm sources of income and capital investment play a major part in the business and livelihood strategies of family farms in the area. They are particularly geared to exploiting

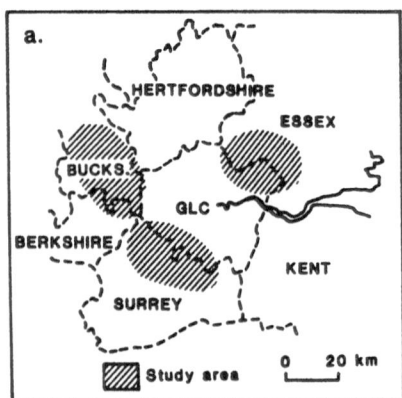

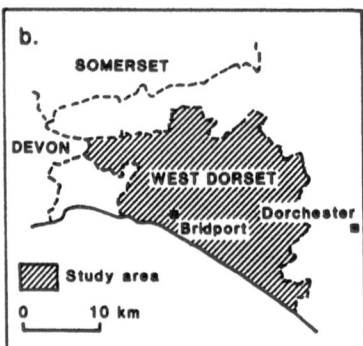

Map 4.1   Geographical location of the study areas

the local surburban consumer market with farm shops, 'pick-your-own' and horse-riding enterprises (Munton *et al.*, 1988).

West Dorset, by contrast, is a predominantly small-scale livestock farming area, with an established specialisation in dairy farming. Agriculture is the dominant land use and the farming community the main purchasers and owners of agricultural land. The county town, Dorchester, is located just beyond the western edge of the study area, and Bridport, a small market town, is the main centre of population within it. The study area is crossed by the A35, a main tourist route to the West Country, and contains part of a National Trust Heritage Coastline. Opportunities for farm tourism, such as bed and breakfast and caravan holidays, are increasing in association with these attractions. The growth in the area's popularity as a retirement location and for second homes and holiday lets has already begun to affect land use and property values around towns and villages and the local social structure.

The farming systems in these two study areas are quite different in terms of their structural organisation and economic complexity. Table 4.1 shows the main characteristics of farms surveyed in the two study areas.

This background information derives from an interview survey of farmers and farm managers conducted for the wider research project with which this work was associated.[7] The survey provided valuable background information on farm enterprises, including their household structure, labour relations, and organisation of asset ownership and management. It provided a base-line sample of 86 farms in the MGB and 99 farms in west Dorset from which to identify women for a postal survey of farm wives, which was undertaken as the first stage of my fieldwork.[8]

Farm typology

The base-line survey of farms in west Dorset and the MGB also provided the basis for constructing a typology of farm enterprises. This typology was originally devised as a means of systematically categorising the extent to which farm enterprises are integrated into the wider 'agroindustrial complex' described in Chapter 2. Individual farms are allocated to one of a matrix of categories, centred around a core set of 'ideal types', representing the extent to which their internal and external relations of production have been commoditised (see Chapter 2). This typological method is described in detail elsewhere (Whatmore *et al.*, 1987a, 1987b). It is adapted here as a basis for examining the relationship between the level of commoditisation in farm production relations and the gender divisions and associated ideologies which structure the organisation of family labour and property. This adaptation centres on the renaming of the

Table 4.1   Selected features of farms in Dorset and the
Metropolitan Green Belt

	FARMSIZE (hectares) (a)	DOMINANT LAND TENURE (b)	MAIN ENTERPRISE (c)	LABOUR (% family labour)	BUSINESS ORGANIS- ATION
MGB	68.8ha	Inhand 38% Mixed 21% Tenanted 41%	Cereals 36% Beef 30% Milk 14% Other 20%	50%	Sole operator 30% Partnership 43% Limited Co. 10% Non-family 17%
DORSET	60.7ha	Inhand 63% Mixed 19% Tenanted 18%	Milk, 60% Beef 12% Sheep 11% Other 17%	71%	Sole operator 57% Partnership 36% Limited Co. 5% Non-family 2%

(a) The median figure has been used in preference to the mean
to reduce the influence of a small number of very large
farm holdings on the distribution.

(b) The term "inhand" is used here to cover owner occupied
holdings in the traditional sense, and freehold land holdings
not directly operated or physically occupied by the owner,
but through some contract or share-farming arrangement.

(c) The figures for both the land tenure of holdings and the
farm enterprises refer to the percentage of farm holdings
whose main/sole type of land tenure or enterprise falls into the
category identified (roughly more than 60% total acreage or
business turnover).

'ideal type' categories to highlight the significance of an underlying
transition in the organising principle of the internal relations of family
farms associated with the commoditisation process; a transition from
'family *labour*' to 'family *capital*' as the level of commoditisation
increases. The 'revised' typology is shown in Figure 4.1.

While at one level this adaptation merely emphasises an aspect of
the commoditisation of farm production relations implicit in the
original version of the typology, it focuses attention more clearly on

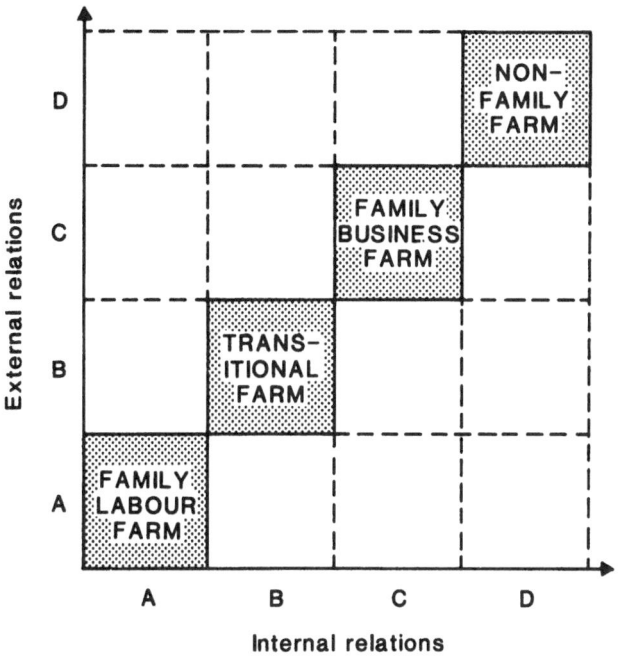

Source: Adapted from Whatmore *et al.* (1987a/b).

Figure 4.1 A relational typology of farm enterprises

the implications of the commoditisation process for the restructuring of gender divisions in 'family farms' and for their class position. Using this typology, differences between the farming systems in west Dorset and the MGB can be further elaborated. Figure 4.2 shows the distribution of farms across the typology for the MGB and Dorset samples. It demonstrates the higher degree of commoditisation of internal farm production relations and the greater diffusion of control over the farm production process away from the family household in the MGB by comparison with the relatively 'simple' structure of farms in Dorset. In particular, it reflects the complexity of land-ownership relations and organisation of capital on MGB farms by comparison with those in the Dorset study area.

Survey of farm wives

In view of the lack of published statistics on farm wives' work, noted in the introduction, the purpose of this survey was twofold: firstly, to

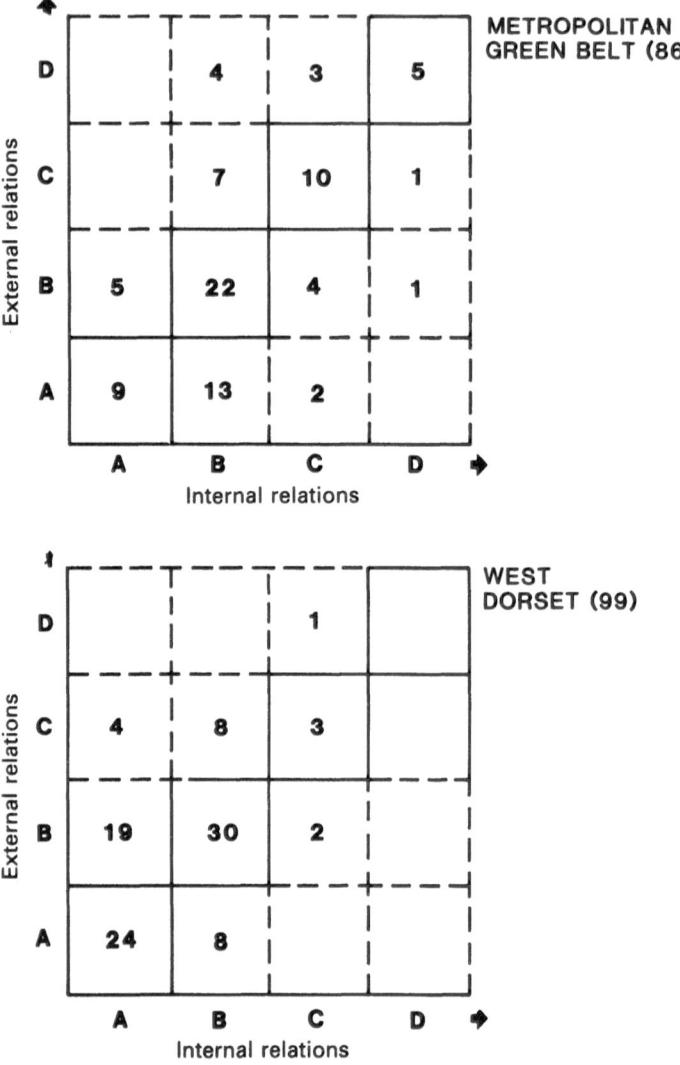

Source: Adapted from Whatmore *et al.* (1987b).

Figure 4.2   Distribution of farms across the typology matrix, by study area

get as representative a coverage as possible of the women involved as wives in family farming; secondly, in the light of the theoretical arguments in Chapters 2 and 3, to reformulate and extend the kinds of questions asked and information gained about gender divisions in the farm labour process and, in particular, about women's domestic labour.

The total sample for the survey of farm wives was 135 women; 75 in Dorset and 60 in the MGB.[9] The decision to conduct the survey through a postal questionnaire was taken on practical grounds, to gain the widest possible information within the time constraints imposed by a full-time job. The problems with this survey method are well known, both for being impersonal and (not unrelated to this) for its low response rate (see Errington, 1985). These problems were certainly not avoided here, but were ameliorated by the fact that I had already met many of the women being asked to participate in the exercise while carrying out the base-line survey of farm enterprises. Indeed it was their enthusiasm for such a project that provided an important motivation for the research in the first place. On the basis of an introductory letter, provision of a stamped addressed envelope and one reminder letter, 81 usable questionnaires were returned out of 135 sent out, or 60 per cent. This was made up of 49 replies in Dorset out of 75 (65 per cent) and 32 replies in the MGB out of 60 (55 per cent).[10]

The information obtained from this survey relates to women on farms across the whole spectrum of family farm categories in the typology, that is family labour farms (category A), transitional farms (category B) and family business farms (category C). The distribution of responses across the typology matrix is shown for each of the study areas in Figure 4.3. In comparing the distribution of these responses across the typology matrix to that of farms in the base-line sample, shown in Figure 4.2 above, a number of points need to be made about the information base for the subsequent analysis of 'farm wives'.

The sample of farm wives is broadly representative of the farm types as a whole, with one notable exception – that of hobby farms located in the AA, AB and BA categories in the MGB.[11] There were 10 such cases in the total MGB sample (none in Dorset) of whom 9 were included in the postal questionnaire sample, and none of whom responded. This means that one type of household/enterprise structure is not represented in the analysis, although the consequences of this omission are not of major significance for the research objectives. The exclusion of non-family farm businesses from those sampled for this exercise also reduced the representation of women in MGB farms at the opposite end of the typology. The responses in Dorset provide a representative coverage of the farm business types characterising the area.

Two further points are worth noting. First, responses across the

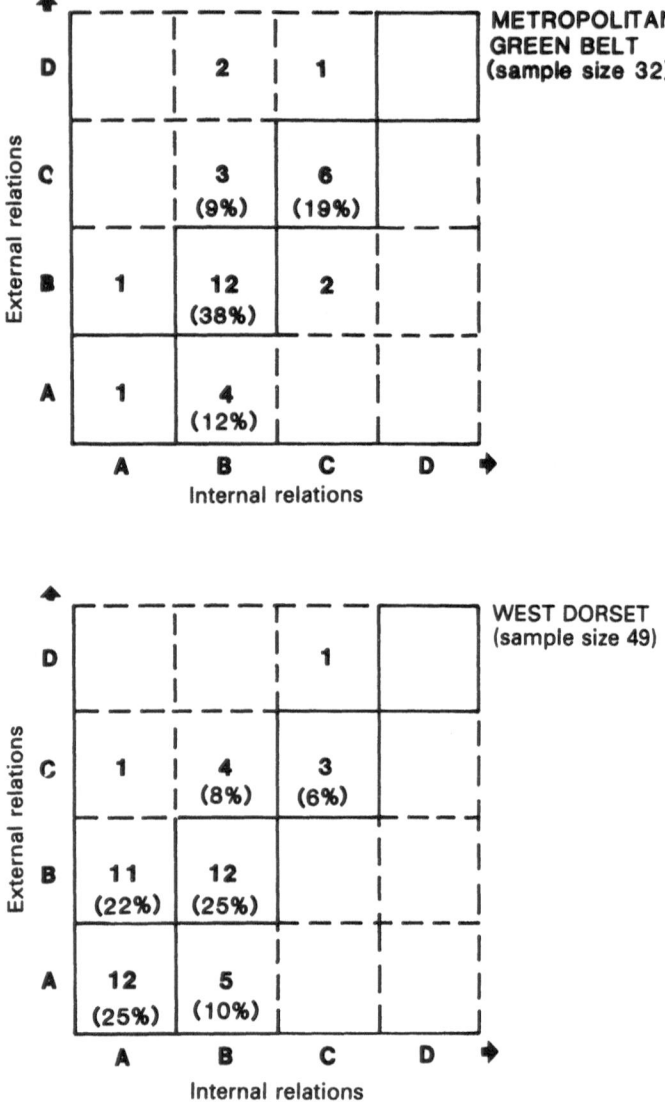

Figure 4.3 Distribution of responses to the survey of 'farm wives', by farm type for each study area

study areas may be affected by husbands' reactions to the question-naire. An indication of this was provided by a letter from one farmer in the MGB, saying that his wife did not wish to take part in the survey. Second, despite saying explicitly in the introductory letter to the survey that women who did not do recognised farm work were of as much interest as those who did, three letters were received in response from women saying that they felt they would 'not be of any interest' as they did not work on the farm. The frequency of farm wives' participation in agricultural labour may therefore be over-represented in the final sample, although the form and conditions of that participation should be unaffected.

The postal questionnaire itself consisted of four main sections: women's social background and current household structure; their labour activities including domestic household labour, agricultural labour, non-agricultural farm labour and off-farm wage labour; their legal and financial interest in farm assets (land and capital) and involvement in farm decision-making and management; and their community activities. In addition to basic questions with regard to these themes, an attempt was made to gauge women's own assess-ments of their positions within, and contributions to, the running of the farm, and farm life (see the appendix). However the experiential aspect of the overall research objectives is more properly dealt with through the ethnographic methods used in the case study exercise (see below). The results from this survey are analysed principally in Chapter 5. However they also inform the detailed domestic political economies of the six case study farms presented in Chapter 7.

The case studies

The case studies focus on the position of women in the domestic political economy of family farms at different levels of commoditisa-tion. Their main purpose is to examine some of the ways in which the 'external' imperatives of the commoditisation process are mediated by and modified through the everyday activities, value systems and work practices of the family household. They seek to elucidate the ideologies mobilised and, in part, produced in the family labour process which legitimise women's exploitation as 'farm wives' through the construction and transformation of gender identities of 'wifehood'. This part of the research seeks to incorporate women's self-understanding into the analysis of how the processes of patri-archy and capitalism intersect in the context of family enterprise.

The intensive research combines the principles of a case study method of analysis with a rather less formalised method of data collection, best described as 'cumulative interviewing'. This combination represents an attempt to adapt some of the ethnographic techniques associated with participant observation to a research context in which participant observation is not itself appropriate, owing to the geographically dispersed nature of the analysis being undertaken. It is the family farm rather than the farming community, as a locally or occupationally defined 'cultural group', that is the focus of analysis. More specifically, it is women's experience and understanding of their position as wives in the farm labour process which the methodology is primarily designed to investigate. This element of the research method is premised upon, and designed to complement, the preceding levels of analysis, rather than to stand in isolation.

The final question in the postal survey asked women if they would be willing to be involved in a more detailed stage of the research. Fourteen women in the MGB sample (45 per cent) gave positive responses, compared to 18 in the Dorset sample (35 per cent). Six of the MGB group were in 'transitional' farm types and eight in 'family business' types. In the Dorset group, ten were in 'family labour' farm types, seven in 'transitional' types and one in a 'family business' farm. It was felt to be important to select case studies from amongst women who positively wanted to be involved in the next stage of the research, as this demanded a high level of personal commitment. The nature of this commitment was explained to women when they were first contacted about the case study exercise.

From an initial seven women participating in the case study exercise, six were eventually included in the final case study analysis.[12] These six case studies include three in Dorset and three in the MGB and are shown in Table 4.2. They cover a range of family farms, at different levels of commoditisation, and women at different points in their life-course and with varied positions and experiences as 'farm wives'. They exhibit a variety of gender relations in terms of the way in which these structure the domestic political economy of the family enterprise. Thus, for example, Sue Price and Hannah Green are extensively involved in different aspects of the agricultural labour process. By contrast, Julie Church, Gayle Brown and Jill Watson are all very active in their own non-agricultural on-farm enterprises and/or off-farm waged work. Vicky Evans, while increasingly called upon as a 'reserve' worker on the farm and having professional qualifications as a radiographer, most readily identifies herself as a 'housewife'.

Table 4.2   The case studies

Alias	Study area	Farm	Farm type	Age
Gayle Brown	MGB	Castleton	B/C	50s
Julie Church	MGB	Naylors Farm	C/C	40s
Vicky Evans	MGB	Rough Farm	C/D	30s
Hannah Green	Dorset	Holly Farm	A/A	60s
Sue Price	Dorset	Fountain Farm	B/B	40s
Jill Watson	Dorset	Vale Farm	A/B	20s

The case study exercise involved a quarterly visit to each farm, centring on a taped 'conversation' with each of the women about their work and position on the farm. This method is termed here 'cumulative interviewing'. It is not a formal technique in the sense of there being an established procedure in the literature to which it conforms, although it can be seen to incorporate a number of aspects of other qualitative research methods.[13] The key characteristics of this method are threefold:

1.   A relationship of trust and familiarity with the research subject was built up over a period of time, in this case one year. Visits develop from a semi-structured interview format to informal conversation. The interview process is thus cumulative, in the sense that each visit presumes the knowledge and interest generated by the previous one and builds upon it. Visits become more an occasion to catch up on family and farm news, to review progress of personal and business problems (money, sick children, exhaustion and so on), than to conduct a formal interview. Although a number of 'topic areas' were identified in advance and covered consistently in each case, the themes which became important were dictated by each woman's own concerns. I rapidly abandoned asking specific questions, partly because the answers could invariably be learnt just by listening, and partly because subjects which would not have been raised, as either unimportant or too sensitive, were brought into the discussion by the women themselves in their own time (see Whyte, 1955, p. 303).

2.   The objective was to get beyond the point of turning up with a tape-recorder and leaving when enough information had been 'extracted'. Visits lasted half a day or a day and extended beyond the taped interview. Conversation and activity preceded and

followed the turning on and off of the tape-recorder. As relationships developed, it was possible to become more involved in farm activities, particularly in helping out with those which fell to women, for example with domestic chores, childcare and meal preparation. This was important because, as the later analysis shows in detail, the working day of farm wives is characterised by the performance of a variety of 'jobs' in response to simultaneous demands made upon them as wife, mother, and 'reserve farm labour'. Helping out with the chores was one way of extending the time I could spend with them and allowed me to see other aspects of their lives than would have been possible with a more formal method.

3.   While the main focus of these case studies was the women themselves, an attempt was also made to involve other family members within a group setting, rather than individually. Bearing in mind that the farmer had already been interviewed independently for the base-line survey, the aim here was to accommodate the interest in the exercise expressed by husbands and adult children, without allowing them to become dominant subjects. In the light of the arguments made earlier in this chapter, this practice also highlighted an important difference in the kinds of information and modes of expression adopted by women on their own, compared to when they were in the company of other family members. The presence of others activated mutual expectations and negotiated ways of relating to one another as a 'couple' or 'family' group. In these circumstances, women fell back on more 'factual' accounts of their work, seeking verification for their account from others.

Visits were set up by telephone a few days in advance, within the quarterly cycle, so that a day could be arranged when I could be sure to see women on their own, even if for only part of the time. It took time to establish that I really was interested in the 'mundane' domestic chores, as well as the 'significant' agricultural work and that chatting was as valuable a way to proceed as my asking a succession of technical questions. This meant discussing the ideas and issues which lay behind the research itself. Exploring our different interpretations of these issues, based on our experiences as women from different generations and backgrounds, itself informed the research. It was also important to be accepted as someone who was interested in, and informed about, farming, because all the women experienced farming as a distinctive 'lifeworld' which set them apart from other people and made it difficult to talk about with outsiders.[14]

The case study material consists of some six hours of taped conversation per case, together with field notes and other contextual material. One consistent element of the enquiry was a 'time diary' exercise incorporated in the recorded conversations which attempted to gauge the format of women's working days, in different seasons. During each visit women were asked to describe their activities on the previous day. This method was used in preference to the more established written time-budget diary because it is more open to women's own interpretation of their 'working day' in which the priorities they choose, and the language they use to describe it, are themselves of significance (see Chapter 6). Moreover women declared themselves to be generally much happier to talk about their activities than to fill in forms (cf. Dixon-Mueller, 1985).

The analysis of case study material works on a very different set of principles from those of survey methods. The validity of the case study as a basis for analysis rests not on its 'representativeness' of a wider population, that is on statistical inference, but on the cogency of the interpretation of the material presented (C. Mitchell, 1983), that is on logical inference. Thus corroboration rather than replication constitutes the appropriate 'test' of the interpretation offered (Sayer, 1984). It is a method which seeks to preserve the unitary character of the research subject (in this case the family farm) while locating it within a wider social context which conditions the collective and individual actions of its constituent members in relation to one another and to the outside world. The cases chosen are therefore neither 'typical', in the statistical sense of an 'average' household or person, nor restricted in their significance to the particular case, but rather 'illustrate aspects of social process and . . . demonstrate certain general theoretical principles' (Wallman, 1984, p. vii).

The analysis of the case studies presented numerous practical problems in terms of the framework within which to structure the volume of material available. The tapes give rise to interpretational problems in that they represent recordings of conversations carried on in the context of other, continuing activities (mealtimes, feeding animals, entertaining children, for example) determined by the demands of the moment. In preference to literal transcription, the analysis of the tapes is based on an interpretation built up from repeated replaying of the recorded conversations, combined with other contextual material. The analysis attempts to reconstruct a picture of the everyday experience of these women as 'farm wives', in which their individuality is retained and their own voice given expression *within* a theoretically informed analysis of the structured negotiation of familial gender relations in the farm labour process. The analysis of this material is presented in Chapters 6 and 7. All the quotations used in these chapters are direct transcriptions from the tapes. However, the next chapter turns first to the survey material to present an analysis of the aggregate pattern of familial gender divisions in the farm labour process.

5 Women's Work and Property

PATRIARCHY OR LIFE-CYCLE?

In this chapter the position of 'farm wives' in the family labour process is examined using women's answers to the survey outlined above. Following the arguments developed in Chapter 3, the four labour circuits used to define the farm labour process are covered, namely domestic household labour, agricultural labour, non-agricultural farm labour and off-farm wage labour. The analysis highlights the inadequacy of traditional explanations of women's position in the gender division of family labour couched in terms of their 'stage in the family life-cycle'. Here women's participation in non-domestic labour circuits is held to vary according to their responsibilities as wives and mothers. Both Marxist and mainstream analyses have recourse to the life-cycle concept. For example, mainstream analyses which recognise the gender division of family labour to be 'asymmetrical' (Symes and Marsden, 1983; Bouquet, 1986) seek explanations for 'variations in women's role [in] family matters such as, age . . . and structure of the household' (Symes and Appleton, 1986, p. 11). In Marxist analysis, the concept has been taken up in efforts to establish the self-sufficiency of simple commodity production in terms of family labour by taking account of its 'variability . . . during the family life-cycle' (Friedmann, 1978b).

The analysis is primarily structured around the categorisation of family farms by their level of commoditisation, described in Figure 4.2. It is argued that the differential pattern of gender divisions in the farm labour process and the varied positions of farm wives exhibited within and between the two study areas is related more closely to the level of commoditisation and to the nature of patri- archal labour relations than to the level of women's domestic responsibilities associated with 'life-cycle stage'.

The analysis is divided into two main sections. The first describes the position of farm wives in the familial gender division of labour on the farm, looking at the division between domestic household labour and agricultural labour and then at the other labour circuits. The second describes the labour relations within which farm wives work.

It focuses on their ownership and control of land and business capital in the conjugal farm, and on their control over the products of their labour. In drawing some conclusions from the analysis, the final section identifies a number of key problems and themes to be taken up through the analysis of the case study material in subsequent chapters.

GENDER DIVISIONS OF FAMILY LABOUR

Domestic household work

The principal axis of the gender division of labour on the family farm is that between domestic household labour and agricultural labour. Domestic household labour is almost exclusively 'women's work' and, in the context of the conjugal household system which pre-dominates, it is the primary area of responsibility and labour activity of the 'farm wife'. It consists of a number of tasks which service the family household, reproducing its labour capacity and social relations of production on a daily and generational basis. These are child-rearing and care, food provision, housework, laundry and shopping (Collins, 1985). In addition, in a limited number of households (five), these labour activities extend to the care of elderly relatives.[1]

The most striking overall characteristic of domestic household labour is the consistency of farm wives' sole, or predominant, responsibility for this work across all age groups, different levels of involvement in agricultural and non-agricultural labour activities, and between farm types and study areas.[2] All farm wives carry out all the main domestic tasks, except childcare. Thirty farm wives (36 per cent) were involved in childcare (covering pre-school and school-age children). Despite the almost identical age profiles of women in the two study areas, a higher number and proportion of women in Dorset (21 and 43 per cent) have childcare responsibilities compared to women in the Metropolitan Green Belt (MGB) (9 and 29 per cent). This is the result of a higher incidence of late child-bearing in Dorset, where there are seven households with children under ten with mothers in the 36–45 age group. There are no such cases in the MGB.

Where there are children in the household it is women who are the primary carers. However to cast an explanation of this gender division of labour in terms of 'stage in the family life-cycle' is to accept as 'natural' a socially constructed division and to take as given

Table 5.1 Women receiving assistance with domestic household labour
(81 farms)

	None	*Occasional*	*Regular*
childcare*	8 (27%)	17 (57%)	7 (23%)
food preparation	48 (59%)	30 (37%)	3 (4%)
housework	56 (69%)	24 (30%)	1 (1%)
laundry	68 (84%)	12 (15%)	1 (1%)
shopping	44 (54%)	31 (38%)	6 (8%)

* Percentage of the 30 households with children.

an aspect of what has to be explained. Attention therefore needs to be focused on the larger pattern of the gender division of domestic household labour. Table 5.1 shows the negligible contribution of other family members to this labour circuit, with the partial exception of childcare.

The pattern and level of assistance differed little between the two study areas, although, with the exception of childcare, more women in Dorset received regular help than in the MGB. This does not reflect a more egalitarian division of labour between husband and wife in Dorset, but is explained by the higher incidence of teenage and adult daughters resident in the household there. In view of traditional arguments that women's agricultural labour participation is a function of their domestic responsibilities, it might be anticipated that paid domestic help is one means of 'liberating' women from these tasks to enable them to have a higher level of participation in other labour activities.

In the farm sample as a whole 27 households (33 per cent) employed wage labour in the performance of domestic household tasks, most commonly housework, and more rarely childcare. However, there is no simple relationship between the presence of paid domestic labour and women's level of participation in agricultural or other labour circuits. In fact a slightly lower number and proportion of women involved in farm work and off-farm work have paid domestic help (16, 50 per cent), than have those with no involvement (18, 56 per cent). While women in the 46–60 and 60+ age ranges are more likely to have paid domestic help than younger women, the employment of domestic wage labour is most clearly related to the income status of the household and class position of the head of household. As many as 62 per cent of family business farms have supplementary paid domestic labour, as against only 32 per cent of transitional farms and 15 per cent of family labour farms.

Agricultural labour

Agricultural labour can be divided into two kinds, administrative work and manual work. Table 5.2 shows the level of farm wives' participation in agricultural labour on the farm, classified into five main task areas traditionally examined in the literature – book-keeping and office work, dealing with telephone and personal enquiries, running errands for the farm, doing manual farm work (livestock related) and harvesting/haymaking work.

Table 5.2   Women's agricultural labour (% women on all farms)

	None	*Emergency only*	*Occasional*	*Regular*	*All*
book-keeping	27	5	14	34	53
	(33%)	(6%)	(17%)	(42%)	(65%)*
enquiries	0	0	19	61	80
	—	—	(24%)	(75%)	(99%)*
errands	12	4	38	26	68
	(15%)	(5%)	(47%)	(32%)	(84%)*
manual work	23	14	17	26	57
	(28%)	(17%)	(21%)	(32%)	(70%)*
harvest/	24	19	15	20	54
haymaking	(30%)	(24%)	(19%)	(25%)	(68%)

'All' refers to the proportion of wives participating in some form in each work category.
*The rows do not always add up to 100%, owing to non-response.

Women can be seen to play a key administrative role which combines the job descriptions of secretary, receptionist, accountant and public relations officer. The work ranges from skilled work, such as book-keeping and secretarial duties, to general 'dogsbody' roles, such as dealing with enquiries and running errands for the business. Most women do these latter tasks, with nearly all women dealing with enquiries, 75 per cent of them on a regular basis, and 84 per cent of women running errands for the business on an 'on-call' basis. Book-keeping is undertaken by 65 per cent of farm wives, 42 per cent on a regular basis. Women are commonly thought not to participate extensively in manual agricultural labour. However the results here show a high level of participation in such activities if the restrictive, official methods of measuring work, such as 'man-hour' equivalents, are abandoned.[4] They also affirm Gasson's (1986) scepticism of Bouquet's contention that farm wives have been 'finally removed from' agricultural labour so that 'it is for them an abstraction because they are not involved in the practice of farm work' (1981, p. 174). A total of 70 per cent of farm wives are involved in some way in manual work on the farm, 32 per cent on a regular basis. Harvesting, by definition a seasonal activity, involves 68 per cent of farm wives, 25 per cent as regular workers.

For all types of agricultural labour, except that of receptionist, a higher number and proportion of women with childcare responsibilities are involved in agricultural labour than those without. Conversely, in terms of childcare and paid domestic help, the only two real 'variables' as far as farm wives' domestic household labour is concerned, only five women with childcare responsibilities do no manual agricultural labour (of whom four also have paid domestic help) whereas women with paid domestic help are the least likely to do manual agricultural work (12, 42 per cent). As with the incidence of paid domestic help, farm wives' overall level and terms of involvement in agricultural work are more strongly related to the extent of commoditisation of the farm enterprise than to the extent of women's domestic responsibilities. The degree of commoditisation of agricultural labour relations is of particular importance in this respect.

Table 5.3 shows the number and proportion of farm wives involved in agricultural labour, the level of their involvement and the types of tasks they perform, in the context of the wider labour relations of the farm. Two patterns stand out. Firstly, a higher proportion of women in solely or predominantly family labour farms are involved in agricultural work, particularly manual and harvest work. Secondly, women's agricultural work on family labour farms is more likely to be performed on a regular, rather than a casual or 'on-call' basis, compared to women in family business farms with a high ratio of hired to family labour. Book-keeping also appears to be a more common activity for women in family labour-dominated farms (particularly transitional farms), as office-related tasks tend to become more 'professionalised', involving farmers themselves and commercial farm secretaries and accountants on family business farms.

Farming Women

Table 5.3 Women's agricultural labour by farm labour type (% of all farms in each category)

		Family labour only	Family labour > hired labour	Hired labour > family labour
book-	none	4 (29%)	14 (32%)	7 (35%)
keeping	reg.	6 (43%)	22 (50%)	6 (30%)
enquiries	none	0	0	0
	reg.	10 (71%)	34 (77%)	17 (85%)
errands	none	1 (7%)	9 (20%)	2 (10%)
	reg.	4 (29%)	15 (34%)	7 (35%)
manual	none	1 (7%)	14 (32%)	7 (35%)
	reg.	8 (57%)	13 (34%)	3 (5%)
harvest	none	0	14 (32%)	7 (35%)
	reg.	6 (43%)	15 (34%)	2 (10%)

Three missing cases, total 78 farms. Rows do not sum to 100% as 'non-regular' cases excluded. Labour categories constructed from the farm typology: see Figure 4.1.

Within both administrative and manual categories, a gender division of labour is apparent in terms of the types of tasks which men and women do, the relations under which they do them (see next section), and the location, or site, of their work on the farm. In terms of tasks, the manual work undertaken by women is restricted to seasonal activities like haymaking and to livestock enterprises and predominantly to non-specialist and secondary farm enterprises.[5] For example, on specialist dairy farms women rarely work regularly at milking the cows, but more usually at calf rearing, or a beef enterprise sideline. Women on arable farms have little if anything to do with the cereal cultivation process and participate only at harvest time, if at all. There also appears to be a territorial dimension to the gender division of labour, with women's work largely restricted to the sites of the farmhouse (domestic household and agricultural administrative work) and the farmyard, except at harvest and haymaking.

Women's agricultural labour is least associated with the technology/machinery-dominated sectors of agricultural production. This represents one of the most consistent of survey findings (see also, Gasson, 1980; Jones and Rosenfeld, 1981) and has a material basis in the lack of formal training or family tuition received by women in technology-related farming skills. Only ten women (12 per cent) in this sample had received formal agricultural training (five in transitional farms, four in family labour farms and only one in a family business farm). However this is a complex chain of casuality to unravel. It is bound up with ideological notions of women's inability to deal with machinery/technology, and of mechanical competence as 'unfeminine', which are not accessible through aggregate survey analysis. Overall, there are relatively few differences in the level and form of women's participation in agricultural labour between the two study areas, except that women in Dorset are more likely to be regularly involved as manual and harvest workers than in the MGB (43 per cent as against 16 per cent). This reflects differences in enterprise structure, labour relations and level of commoditisation between the two areas. These patterns resemble quite closely the regional and sectoral picture presented by official statistics (Gasson, 1989).[6]

In the context of the predominantly responsive and 'on-call' nature of women's agricultural labour, it is important to gauge the extent to which they have primary responsibility for any specific agricultural tasks. This represents work conditions in which women have a high degree of self-determination and in which the end products are most tangibly the results of their own labour, even if their subsequent control over any income generated from them is limited (see the next section and Chapter 7). Farm wives' *primary* responsibility for specific agricultural tasks is infrequent in terms of the production of major agricultural commodities.[7] A higher proportion of farm wives in family labour farms (69 per cent) have primary responsibility for some area of farm work than in the more commoditised farm types (57 per cent in transitional farms and 54 per cent in family business farms). This is most likely to be livestock rearing/feeding in the case of family labour farms; and book-keeping and office work in transitional farms.

Other kinds of work

Just as there is no simple association between farm wives' domestic household labour responsibilities and their participation in agri-cultural labour, so their involvement in non-agricultural farm labour and off-farm waged work bears a complex relationship to the primary axis of the gender division of labour on the farm. Table 5.4 shows that in terms of non-agricultural farm labour (i.e. those tasks or enter-prises not directly related to the cultivation or rearing of plants and

Table 5.4   Women's non-agricultural farm labour

	All farms		MGB		Dorset	
	none	reg.	none	reg.	none	reg.
farm shop/PYO	46	20	12	13	34	7
	(57%)	(27%)	(38%)	(47%)	(69%)	(14%)
bed and breakfast	68	19	30	1	38	8
	(85%)	(11%)	(97%)	(3%)	(77%)	(16%)
horses	69	6	26	4	43	2
	(87%)	(7%)	(84%)	(13%)	(88%)	(4%)

Three missing cases, total 78 farms.

animals) women are actively involved in these study areas in farm shops and pick-your-own schemes, bed and breakfast and tourist accommodation, and commercial horse-riding activities.

These kinds of jobs can be seen, in the first two cases at least, to be commoditised forms of women's domestic household labour. They extend the production of food and consumption goods for the family household to a commercial market where they are realised as an exchange value (see also Bouquet, 1986, on the bed and breakfast issue). Clearly, farm wives exploit different consumer markets in the two study areas. In Dorset, the tourist trade is the main source of income for both bed and breakfast accommodation (and renting out farm cottages for self-catering holiday lets), and horse-riding activities. Their work in this labour circuit is highly seasonal. In the MGB, by contrast, it is the local resident suburban consumer market, willing and able to pay a premium for fresh farm produce and/or a 'day out in the country', who form the base for the farm shop and pick-your-own activities. It is the children of this resident population who form the staple market for the commercial horse-riding activities.

Of all the labour circuits, off-farm waged labour is the one in which farm wives participate the least, with over 85 per cent of women having no paid work off the farm. Amongst those who do, there is a minority associated with family business farms who pursue an independent career or job. The majority of cases, associated with family labour and transitional category farms, are women who work off-farm in order to contribute to the family household budget, where the farm business itself produces insufficient income to meet household needs. These women are concentrated in the secretarial and clerical sector and tend to work part-time. This survey suggests that, again, there is no simple relationship between women's off-farm labour and their domestic household responsibilities or farm labour. What is striking is that a much larger number of farm women held paid jobs until they married, in contrast to the more usual contemporary pattern of giving up employment on the birth of a first child.[8] Only 30 per cent of farm wives had never held a paid job, with a slightly higher proportion in Dorset. A total of 64 per cent of them gave up their jobs on marriage, coinciding with moving to the farm, and only 20 per cent on the birth of a first child. Moreover those least involved in agricultural work were also the least likely to hold off-farm jobs, and to conform most closely to the stereotypical 'housewife'. This is particularly the case in the MGB, associated with a higher incidence of family business farms.[9]

This section suggests that the relationship between farm wives' domestic household labour and their involvement in agricultural, or other, labour activities is more complex than the notion of a 'compensatory' mechanism or balancing act assumed in many analyses. Most significantly, domestic household labour is a ubiquitous feature of women's workload and, even in the case of childcare, bears little relation to variations in the level of their involvement in other labour circuits.

PATRIARCHAL LABOUR RELATIONS

The relations within which women participate in the farm labour process share certain common attributes which distinguish their conditions of work from those of their husbands and sons. As Bouquet has argued, women bear a highly ambiguous relationship to the kinship structure of the family farm, whereas

men usually inherit the farm, ... women usually marry in ... While there is a certain continuity for men throughout their lives, there is a degree of disjunction for women in that they are born into one family and marry into another. (Bouquet, 1984b, p. 73)

In consequence, women's rights over property reflect, and reinforce, social constraints on their independence arising from patriarchal kinship relations and ideologies which subsume women within the identity and material status of their husbands (Hirschon, 1984). However, as the analysis below demonstrates, a series of potential contradictions exists between the position of women as wives in the farm labour process, organised around the conjugal household unit, and their position in the patrilineal kinship system that structures the control and transfer of the means of production between father and son(s).

Gender divisions of land and capital

Farm wives' interest in farmland is limited in extent, with only 21 women (26 per cent) holding an interest in any form. It is more widespread in the case of farm capital with 51 women (63 per cent) holding an interest of some kind. This reflects the significance of land in the social relations of farming, as a principal means of production and as symbolic capital, which distinguishes it from other forms of capital. Land is the principal asset in the majority of farm businesses,[10] giving them a distinctive equity profile against other sectors of industry and an unusually high collateral base for borrowing purposes, in relation to their capital or business asset structure (Traill, 1980). As a result, control over the land carries with it leverage over other forms of capital. Control over land thus has particularly intense associations in family farming.

In the vast majority of cases where women do hold an interest in land or capital (15, or 70 per cent for land and 42, or 80 per cent for capital), it takes the form of 'joint ownership' with their husbands, that is, it is not an individual right but conditional upon, and structured by, their conjugal status. These forms of ownership often represent tax avoidance measures whereby wives are incorporated into the legal title to land or capital to reduce taxation liabilities on asset transfer, should the husband die (Nix, 1987). In this way, women act as a channel for securing the transference of the means of production intact to the next (male) generation. The period over which they retain the legal title may be extensive, but rarely is it retained by them for their own use or disposal. This practice is commonest amongst older married couples, often on the suggestion of legal advisors, where the threat of capital taxation is more pressing. The survey showed women in the '45–60' and 'over-60' age groups to be the most likely to have an interest in land in joint ownership with their husbands (66 per cent of such cases), and the most likely to be partners (75 per cent) in the ownership of business capital. This supports the limited evidence available from other studies (Salamon and Kiem, 1979; Geisler *et al.*, 1985) which suggests that women's *sole* ownership of farm land is largely restricted to widowhood.

Table 5.5   Women's ownership of land, by farm type (all farms)

	None	Own it	Own part	Own with husband	Other	NI
A	20 (77%)	0	0	5 (19%)	0	1
B	26 (62%)	1 (2%)	3 (7%)	9 (21%)	1 (2%)	2
C	11 (85%)	0	0	1 (8%)	1 (8%)	3

NI   no information provided. A is family labour farm, B is transitional farm and C is family business farm.

As Table 5.5 shows, it is women in the transitional farm category (B) who are the most likely to have an interest in land, with 14 such cases (32 per cent), against only two (14 per cent) and five (19 per cent) in family business (C) and family labour (A) farms respectively.

A similar pattern is shown in Table 5.6, with partnership arrangements predominating as the commonest form of wives' interest in farm capital. The level of involvement of women as partners in the business is highest (64 per cent) in 'transitional' farms and markedly greater than in family labour farms (24 per cent) and family business farms (12 per cent).

Table 5.6   Women's ownership of capital, by farm type

	None	Partner	Co-director/ shareholder	NI
A	12 (46%)	10 (38%)	0	1
B	11 (26%)	27 (64%)	3 (7%)	0
C	6 (46%)	5 (38%)	2 (15%)	0
all	29 (36%)	42 (52%)	5 (6%)	1

Rows do not sum, owing to four cases of missing data.

In describing women's legal interest in farm assets, account needs to be taken of the extent to which the 'simple' form of patriarchal ownership, where all rights are vested in the husband/father, has been modified by the formal inclusion of other, predominantly male family members – most commonly his sons, brothers or father – into more complex forms of patriarchal property ownership. This survey indicates that women are more likely to be involved across all forms of corporate family ownership where a son is also involved in the business, particularly where the son is an active working partner and decision-maker but does not as yet hold a legal stake in the business. This supports the suggestion, made above, that women's legal interest in capital assets, like land assets, represents a 'channelling' mechanism for transferring property rights to the next male heir, rather than a basis for their independent interest in the use and disposal of these assets.

As with women's joint ownership of land assets, the inclusion of women as formal business partners or company directors does not necessarily coincide with any increase in their control over the use of those assets in practice, although it does compromise the integrity of the farm unit. While farm wives' legal rights in land and business assets secure tax benefits for the business and cement the incorporation of women's inherited or personal capital into the business (see below), it gives them *de jure* rights over the disposal of those assets which *potentially* enable women to alienate them from the conjugal farm. The exercise of such rights by women, for example, on divorce, or in efforts to realise the capital value of land rights as a source of independent income, holds serious consequences for the financial viability of the farm enterprise and for the reproduction of patriarchal control over the means of production through patrilineal kinship practices.

This represents a key contradiction between the patriarchal gender relations, inscribed in bourgeois familial ideology, and the commoditisation and individualisation of social relations under capitalism within which women have gained independent legal status and property rights in productive assets. As women renegotiate their role on the farm and in the family business, in the context of their increasing *de jure* rights over the means of production, the ideological apparatus surrounding the gender relations of property ownership will be reshaped. Currently, however, these contradictions remain latent and, in the context of social disapprobation experienced by women exercising their rights over farm assets in an independent way, farm wives' rights over land and capital remain largely unrealised in terms of their increased control over the production process (see Chapter 7).

One consequence of this distinction between *de jure* and *de facto* rights is that the interpretation of variations in the incidence of farm wives' holding an interest in the land between farms at different levels of commoditisation is highly problematic. In the cases of both land and capital, women's legal interest in farm assets is most commonly associated with transitional farms. Whether this is for accounting and taxation reasons related to the development of more complex forms of business organisation, or whether it reflects a higher level of inherited capital assets invested by women in such farms is difficult to determine. There is no simple relationship, however, between the level of personal capital or land invested by women in the conjugal farm and their legal status or financial interest in farm assets and property.

Sixteen women were the recipients of bequests and gifts of land across the whole sample.[11] Twelve of these cases were in Dorset and only four in the MGB. This is in part reflected in the higher proportion of women in Dorset (33 per cent) than in the MGB (15 per cent) with an interest in land. The relationship between the incidence of women bringing land to a farm through marriage and their holding a legal interest in the land is not straightforward. Of the 21 women who hold an interest in land on the conjugal farm, only eight had been given or bequeathed land; the rest held an interest jointly with their husbands. In these cases women had no rights individually over the land until their husband's death, and not necessarily then. Moreover women who have been bequeathed land which they have invested in some way into the conjugal farm do not necessarily retain any legal rights over its use or disposal. The predominant pattern of disposal of land inherited by women in their own right is that of integrating it into the conjugal farm (10 cases, 59 per cent), or selling it and investing the capital raised into the farm (two cases, 12 per cent).

Those few cases of women who are outright owners of all or part of the land on the farm (three) all had land bequeathed to them by their deceased fathers. Significantly, while few in number, all these women were in transitional category farms. Moreover farm wives in 'transitional' farms were the most likely to be recipients of land bequests. In combination, these findings suggest that the incorporation of land inherited by women into such farms may be material to the process of expansion and development from a more marginal production capacity. The numbers involved are too small to support this as more than a suggestion here, but it is pursued further in Chapter 7 through the analysis of the case study material. What is not in doubt is that, in all cases, women's inheritance of land has directly benefited the conjugal farm and, in most cases, her husband's property assets.

The incidence of women inheriting capital assets (usually from their father) and integrating them into the conjugal farm business is lower than is the case for land. Only 11 women had received bequests or gifts of capital assets across the whole sample, seven in Dorset and four in the MGB.[12] Unlike the situation for land, women who invest inherited capital assets into the conjugal farm are likely to hold a legal interest in the ownership and control of the farm business. However, where farm wives do hold an interest in capital or land in the conjugal farm, this is usually restricted to a minority stake as a junior partner and does not necessarily signify a commensurate status in the decision-making and management processes through which these assets are used (see below).

While the inheritance and investment of land and capital assets by women into the conjugal farm is clearly related to whether or not they come from a farming background themselves, social background has little bearing on whether or not wives hold an interest in the farm land or business. A complicating factor is the relatively high number and proportion of farm wives who have invested income of their own into the conjugal farm. As Table 5.7 shows, such income is drawn either from their current or past earned income, from personal savings or investments, or from inherited money.

Table 5.7   Women's independent income, by farm type

	None	Pension/ life assurance	Money from parents	Investment income/ savings	Other	All
A	12 (46%)	0 —	1 (4%)	7 (27%)	6 (23%)	26
B	20 (48%)	4 (10%)	2 (5%)	12 (29%)	2 (5%)	42*
C	7 (54%)	0 —	0 —	5 (38%)	1 (8%)	13
all	39 (47%)	4 (5%)	3 (4%)	24 (30%)	9 (11%)	81*

*Includes two cases with no details provided.

While women in more commoditised categories of farm are more likely to have private sources of income, they are less likely to invest any of it in the farm business. Seven women in each of the family labour and transitional farm categories have invested private income in the conjugal farm, all of them, bar one, in the Dorset study area. Only one woman in the family business category (in the MGB) had done this. These cases represent 88 per cent of all women in family labour farms who have private income, 32 per cent in transitional farms and 14 per cent in family business farms.

The relationship between the extent of family members' labour and investment in the farm business and their legal stake in it, the traditional justification for the inclusion of sons as partners in the farm business (see, for example, Nalson, 1968; Marsden, 1984), exhibits a further dimension of gender inequality in family labour relations. Two indicators of the agricultural labour activities performed by farm wives – book-keeping and manual labour – together with women's assessments of their labour contribution are examined against their legal status in the farm business in Table 5.8.

Table 5.8   Women's ownership of capital in relation to their agricultural labour

	No status	*Partner*	*Company director*	*NI*	*All**
book-keeping	17 (32%)	31 (58%)	3 (6%)	2	53 (65%)
manual labour	19 (33%)	30 (52%)	4 (7%)	4	57 (70%)
> 1/4 of family labour	13 (30%)	16 (36%)	2 (6%)	8	44 (54%)

*'All' refers to all women undertaking agricultural labour as measured by these three criteria as a percentage of the total sample.

Only 36 per cent of those women who assess their contribution to family farm labour at more than a quarter had partnership status in the business, only just higher than the proportion and number who had no business status. Moreover 19 women, 52 per cent of those who assessed their farm labour contribution as less than one-quarter of total family labour, were also partners in the business, against 35 per cent who held no position. The proportion of women doing manual and book-keeping work who hold a partnership status in the business is just over 50 per cent. However 41 per cent of women who did no book-keeping and 52 per cent of those who did no manual labour on the farm were also partners in the business. In other words, women are not necessarily more likely to gain a legal stake or share in the control over the farm business whether they contribute a large amount of agricultural labour or none at all.

Decision-making and remuneration

The level of farm wives' involvement in management and decision-making processes on the farm might also be expected to be related to either their farm labour contribution or their financial and legal stake in the land and business assets of the farm, or both. Table 5.9 shows the extent of women's participation in farm decision-making processes by farm type.[13]

Table 5.9   Women's participation in decision-making, by farm type

		Never	*Emergencies only*	*Occasionally*	*Regularly*	*NI*
A	1	7 (27%)	3 (12%)	9 (35%)	6 (23%)	1
	2	9 (35%)	2 (8%)	8 (31%)	6 (23%)	1
B	1	13 (31%)	2 (4%)	16 (38%)	11 (26%)	0
	2	10 (24%)	1 (3%)	18 (43%)	13 (31%)	0
C	1	6 (46%)	3 (23%)	3 (23%)	1 (8%)	0
	2	4 (31%)	1 (8%)	5 (39%)	3 (23%)	0
all	1	26 (32%)	8 (10%)	28 (35%)	18 (22%)	1
	2	23 (28%)	4 (5%)	31 (38%)	22 (27%)	1

1: Day-to-day management decisions. 2: Long-term management decisions.

Overall there is a slightly higher rate of participation in long-term decision-making than in day-to-day management decisions. A higher proportion of women in family business farms are likely to be completely excluded from farm decision-making and management processes, particularly on a day-to-day basis (46 per cent), than in transitional (31 per cent) or family labour farms (27 per cent). They are also less likely to be involved in decision-making on a regular basis than women working in the context of less commoditised production relations (8 per cent against 23 per cent and 26 per cent respectively). The pattern of involvement is almost reversed between the two study areas, with more women involved in day-to-day decision-making in the MGB than in long-term decisions, and the opposite applying in Dorset.[14] Moreover, while it is slightly more likely for women with a formal business status to be actively involved in the decision-making process, this is not necessarily the case. For example, 46 per cent of women who have partnership status declared that they never participated in decision-making.

The relationship between women's involvement in decision-making, asset ownership and the commodity-producing labour processes and their receiving pay is a complex one. It is bound up with the extent to which monetary measures of the value of work associated with capitalist society have been internalised within the family household more widely, itself related to the level of commoditisation of the farm. In the family labour farm, where the labour of all family members is uncommoditised, it is less the lack of pay that distinguishes farm wives' labour than the lack of control over and responsibility for any specific product which they participate in producing. In the case of family business farms, with more commoditised labour relations, it is the issue of payment which is a more significant feature. However the question of pay is a complex one, reflecting not only the internal labour relations of the farm business but also external influences, such as the state's treatment of different forms of business organisation for taxation purposes.[15]

A total of 46 farm wives in the sample (57 per cent) receive payment in some form for their agricultural labour. As Table 5.10 shows, the proportion of women receiving pay for their non-domestic farm labour activities rises with the level of commoditisation of the relations of production on the farm. This is not as contradictory as it might appear. Although significantly fewer women in family business farms are likely to be involved in farm labour than in family labour and transitional category farms, they are more likely to receive payment for their labour when they do participate. Given the distribution of farm types between the study areas, it is more likely for farm wives in the MGB to be paid than in Dorset. In addition the incidence of payment is related to farm type in terms of business organisation, rising with the incidence of more corporate business forms involving multiple family members, including the farm wife. Moreover the form of business organisation also appears be linked to the form of payment received by those women who receive pay of any kind.

Table 5.10   Women receiving pay, by farm type (all farms)

	No pay	*Pay*
family labour farms (A)	11 (42%)	15 (58%)
transitional farms (B)	20 (48%)	22 (52%)
family business farms (C)	4 (31%)	9 (69%)
all farms	35 (43%)	46 (57%)

Women who are a nominal or active partner or director of a family farm business are the most likely to receive payment in the form of a wage, often a 'token wage' which is counted against the business's tax liabilities. Such corporate forms of family farm business are more common in the MGB than in Dorset. In the context of non-commoditised family labour relations, women are more likely to receive payment in a more general form of a division of the money income from the sale of farm produce, or 'share of the profits'.

This section has highlighted three dimensions of patriarchal labour relations in the family farm: firstly, the limited extent of women's legal and financial interest in the means of production; secondly, their restricted involvement in the farm decision-making process; and thirdly, their lack of control over the products of their labour and, particularly, in a commodity production context, over the money income generated from it.

PATTERNS AND PROBLEMS

Traditional explanations of the pattern of wives' involvement in farm labour, in terms of their primary responsibility for domestic household labour associated with stage in the life-cycle, have been shown to be not only tautologous but inconsistent with the evidence presented here. On the one hand, women's responsibility for, and work in, this labour circuit is more or less ubiquitous, thus carrying little weight as an 'explanatory variable'. On the other hand, for those tasks which are not common to all women, most significantly childcare, no consistent relationship is found between women doing such work and their active participation in other labour circuits.

Equally, functionalist explanations of the gender division of family labour, as being necessary to the survival of the family farm (Bouquet, 1984a; Flora, 1981), fail to explain its gendered structure, that is, why it is women as opposed to men who do domestic household labour. What stands out most from this survey is the far lower variability in men's participation in domestic household labour than in women's participation in other labour circuits. Two features are of particular note. First, that whatever else women do on the farm it is in *addition to*, rather than *instead of*, their household tasks and responsibilities and, second, that whatever the level of farm wives' other labour activities there is little variation in the degree to which domestic household labour tasks are shared with husbands or other family members.[16]

The analysis also demonstrates that the pattern of farm wives' participation in these other labour circuits is more consistently related to the degree of commoditisation of the wider relations of production on the farm than to their 'domestic circumstances' or personal characteristics, such as age or social background. Here again, this contrasts markedly with a lack of variation in men's participation in domestic household labour in relation to commoditisation. It indicates that commoditisation results less in the outright exclusion of women from agricultural labour, as is commonly suggested, than in changes in the conditions of women's participation and, particularly, the casualisation of their labour and loss of self-determination over particular products or activities. This process of casualisation may then itself be partly responsible for the 'disappearance' of farm wives from 'visible' labour circuits in more commoditised farms becuse it results in wives' work falling outside the standard measures of labour-time.

However the relationship between commoditisation and women's labour participation is not straightforward. While commoditisation as measured by the composite typology index presents a more consistent relationship than traditional correlations with particular morphological features of farm businesses, there are two problems associated with this index. Firstly, it raises interrelated issues of social class, enterprise mix, locality and such like, which cannot be easily disentangled at the level of aggregate analysis. Secondly, while correlations betwcen patterns of commoditisation and women's work may indicate useful avenues to pursue in terms of seeking causal explanations, they cannot advance analysis vcry far down them.

The conditions or terms of women's participation in the farm labour process are characterised across all labour circuits by a number of features which distinguish them from those of men. The most important of these in terms of this analysis is their lack of control over the means of production and the products of their labour. These patriarchal labour relations are structured through the interlocking of conjugal kinship relations, as the organising principle of family labour, and patrilineal kinship practices which organise the ownership and transfer of property rights. The implication is that the division of labour tasks is only one manifestation of processes of production, consumption and distribution on the family farm which are fundamentally gendered.

Two particular dimensions of this gender structuring are notable in this analysis. Firstly, women's rights in land and business assets and involvement in the business decision-making process are principally structured by their *specific* gender position as 'wives'. Gender inequalities in respect of control over the means of production are realised through the terms of the marriage contract whose specific concrete form and ideological connotations are reshaped in the commoditisation process. Secondly, women's rights over land and capital assets are mediated by their kin relation to men more generally, especially to fathers and sons. Their incorporation into corporate forms of patriarchal property ownership is principally a means of channelling assets between male kin along patrilineal lines.

Although the small number and proportion of women with property rights in the conjugal farm is itself of significance, the analysis of those women with rights over farm land or capital is compromised by the small numbers in the sample. Nevertheless it is consistent in suggesting that, while less dramatic than the dowry systems described in studies of more 'exotic' cultures, the incorporation of women's inherited land and capital assets, and independent income, into the conjugal farm is material to the 'survival' of individual farm businesses in the commoditisation process and to the restructuring of the social relations of domestic commodity production in general. At the same time, however, the analysis indicates some areas of contradiction emerging as a result of changes in women's property status as individuals and the ethos of the patriarchal family and associated ideologies of 'wifehood' which subsume women within the identity and material status of their husbands.

The survey results repeatedly raise questions about the way the massive gender inequalities identified within the family labour process are legitimised. How are divisions in labour tasks, property ownership and farm product *experienced* by the women (and men) participating in the labour process? What do these differences in labour experience *mean* to those involved and how are these meanings constructed? Central, in terms of the analysis presented here, is the way particular tasks and conditions of work are gendered so that women's work both builds on and reinforces their subordinate status through the institution of marriage and the gender identity of wifehood. The analysis at this aggregate level has been able no more than to hint at some of the ways in which gender ideologies are actively mobilised in the labour practices of family farming, such as through territorially and technologically sited power relations. This problem brings to the fore the more general limitations of extensive analysis in getting to grips with the processes which underlie the patterns of gender division and wives' work described in this chapter. The next two chapters turn to a more intensive level of analysis, using case study material, to address these questions.

6 'Being the Farmer's Wife'

IDEOLOGY AND THE LABOUR PROCESS

This chapter examines women's experience of the farm labour process, focusing on the ideological construction of gender identities of 'wifehood', their mobilisation in the work practices of farm families and their role in legitimising the patriarchal labour relations described in the last chapter. Within the theoretical framework set out in Chapter 3, the analysis takes an argument from Burawoy (1979, p. 30) as its starting-point, namely that the labour process must be understood in terms of the specific combinations of coercion and consent that elicit the co-operation of those exploited by the relations and practices which define it. Applying the methodological arguments set out in Chapter 4, the analysis seeks to inform a theory of patriarchal labour relations from women's own understanding of their everyday experience of the farm labour process as 'wives'. It draws on taped conversations with women in the six case study farms introduced in Table 4.2. All the family names, farm names and place names used in the analysis are aliases to preserve the anonymity of those involved.

The analysis presented does not deal with the complex processes responsible for producing the wider value systems and gender ideologies of the local and farming communities in which these case studies are located. While this undoubtedly constrains any attempt to reconstruct the 'lifeworlds' of the women at the centre of the case studies, the objective here is less ambitious. It is to show some of the ways in which the patriarchal gender ideologies associated with the family labour process are reproduced through women's own subjective experience and taken up in the discursive positions which they adopt (Weedon, 1987). A number of points of contradiction are suggested between patriarchal gender ideologies and the process of commoditisation, which are experienced by women as tensions between their 'roles' as wives and their identity as individuals.

The first section translates some of the general characteristics of the gender division of labour, described in the last chapter, into shared features of women's everyday experience as 'farm wives' across all farm types. The second section distinguishes two ideologies of wifehood associated with different levels of commoditisation and explores the different ways in which these ideologies are constructed and mobilised in the labour practices of farm families. Finally, some conclusions are made on the basis of the analysis presented and a number of difficulties with it are considered.

FAMILY TIES AND WORKING LIVES

Many women responding to the questionnaire made the point that their most important role on the farm was quite simply to be there when needed and to be able and willing to turn their hand to anything, depending on the priorities set by their husband. This defined the role of wifehood for many women. As one respondent put it, her most important role was simply 'being the farmer's *wife!*' (original emphasis). Thus, while individual women's experience of the farm labour process as wives is diverse, a number of features unite their experience across a range of structural conditions and concrete circumstances. Two, interrelated themes are addressed here which emerged from the recorded conversations with women on the six case study farms as common to their experience as farm wives; these are 'the working day' and 'women's work'.

The working day

In Chapter 5 it was argued from the analysis of survey material that women's experience of work as farm wives had three distinctive features: first, that their primary, or sole, responsibility for domestic household labour is pivotal to their wider pattern of work; second, that their work is largely responsive rather than initiatory or self-determined; and third, that it is characterised by the simultaneous performance of several tasks associated with women's multiple roles as wife, mother and reserve farm labour.[1] Vicky Evans defines the experience of these working conditions by contrasting it to the pattern of her husband's work experience: 'He has more spare time than I do at certain times of year, as a housewife's job is non-stop whereas his is very much go and then stop a bit.'

A comparison of the working day of women in the very different circumstances of a traditional 'family labour farm', a 'family business farm', and in transitional cases where one or other partner is engaged in off-farm waged work, illustrates the diversity of compromises built around these three features in structuring the experience of 'farm wives'.

Hannah Green, Holly Farm, Dorset (farm type A/A)

Up at 7.30, put the kettle on and make tea, go out to fetch the cows in for morning milking, mix calf milk and feed the calves, take a cup of tea out to Tom [in the milking parlour]. Feed the yearlings in the yard and bed out the horses and feed them. Feed the poultry. Get the breakfast [for her husband, herself and usually her daughter] for about 9.00. Clear away, do some washing, think about what to make for lunch. Make lunch for about 1.00. Clear up, hang out the washing. Go out to help shift the sheep, I'm used as a stationary gate. Back to the house, make the beds and do the bathroom then its really whatever turns up, there's always some 'rep' [sales representative for an inputs firm] turning up or something and I'm the first line of defence, it's my job to keep them away from Tom [husband] unless he's said otherwise. Make the fires up in time for everyone coming in to tea. 4.30 have tea, usually make a cake, everyone comes in for that [usually husband, daughter and son-in-law]. After tea I fetch the cows in again for afternoon milking [daughter does the milking] and feed the calves and bed the horses. Make tea for about 6.00 [supper]. Clear up and then go and sit by the fire in the back room until bed, about 9.30.

Julie Church, Naylors Farm, MGB (farm type C/C)

Up at 7.00, got breakfast for everyone and children ready for school, put the washing on, took the children to the station for school about 8.00. Came home and did the essentials – making the beds and putting the washing in the dryer. Loaded up the car with stock [dried flowers] and went to see some buyers I'd fixed up the day before. I had also arranged to go to see some new seed suppliers I'd heard about in Halstead and my regular one in Suffolk. Spent the day seeing them, bought some bits and pieces for making up displays, but had to get back in time to pick the children up from the station at 4.30; I just made it. Then went out again to see my father, who's been very ill, in hospital, back about 7.00. Went out to the garden to pick some vegetables for the evening meal, cook for the family at about 8.30, wash up, clear away a bit, then I had to do some ironing as there was nothing left for school the next day. About 10.00 I had a go at my father's cattle records [registering pedigree cattle] until bedtime at about 11.30.

Jill Watson, Vale Farm, Dorset (farm type A/B)

Get up about 7.30, dress the children, Duncan [her husband] helps get the children's breakfast while I do some hoovering. Left the house about 8.15, took son to playschool and then on to work [school] for about 9.00. [Mother-in-law comes in to look after baby daughter and cook lunch for husband.] I had two free periods in the morning so I planned my next couple of domestic science lessons. We had a staff meeting at lunch time, then after lunch I took netball for the afternoon. Got back home at about 5.00. [Duncan picks up son using mother's car.] Played with the children for a bit, made their tea and got them to bed and then start cooking the supper when Duncan comes in, about 7.30ish. In the evening I usually do babygear things, stocktaking or accounts or something [there's a lot of washing and ironing and mending to be done for the second hand clothes]. I do that in front of the TV, then get to bed between 11.00 and 12.00 usually.

Vicky Evans, Rough Farm, MGB (farm type C/D)

Up at 6.15, walk the dog, on the way back feed the calves and straw up the calves and horses, bagged up some feed, feed the ewes in the barn and let the hens out. 7.00 get the children up, washed and dressed for school and make breakfast for the family. Take the eldest child to school for about 8.30, take the other one with me to Linfold to take the dog to the vet, on the way back buy some henfood. Back on the farm by about 9.15, pen the sheep and inject them against 'pulpy kidney'. Check their feet, only a couple need doing but I have to go back home to get the clippers. Back home at about 9.45, have a wash, out to a friend's for coffee [with youngest child]. Check in with husband on the way, at the garage, to check if there's anything he wants doing. On way back from friend's house go to the bank for Richard [husband] put the two ewes that need their feet clipped in the back of the car to take back to the yard. 1pm, make lunch, then take down the Christmas decorations. About 3.00 go back to Linfold to collect the dog and pick up son from school. Back to the farm. Feed the calves, check the soweys [a rare breed of sheep] and feed the poultry again. About 4.30 back home, light the fires and get the children's tea, get them to bed by about 6.30. Make supper for Richard for about 7.00. Wash up, clear up the kitchen and then from about 9.00 I sat in front of the telly and knitted a pair of gloves that I'd promised Adrian [eldest son] for school. Bed about 12.15.

The process of combining this diverse range of labour tasks and of co-ordinating different roles is not a conscious activity but a habitual experience arising from daily practice. This was amply demonstrated during the course of the research itself, a series of interruptions from the telephone, callers, and a husband or children requiring attention is recorded on every tape.[2] As a result the exercise of trying to recall the activities of a 'working day' was difficult for these women because, as Sue Price put it,

> ... you see, you develop the knack of being able to do two or three things at once. I mean most people don't realise that, ... that's why when you sit down and talk about it its very hard to separate.

In recounting the events and activities in a working day, domestic household tasks are the least likely to be recalled of the two or three 'things' being done simultaneously, except for meal times when the family household is united around the domestic event. This draws a close parallel between the methodological and analytical difficulties arising in trying to assess women's labour on the farm, discussed in Chapter 4, and women's own experiences of the farm labour process. It also highlights a second common thread in farm wives' experience of work, centred on the contradictions arising from an ideology which accords women primary responsibility for domestic household tasks but at the same time takes for granted and undervalues the work and skills involved.

'Women's work'

While domestic household labour was a central feature of all the women's working lives, it was so much taken for granted that the labour process itself and the tasks involved disappear from their conscious actions, as well as from the consciousness of other family members. Julie Church best conveys the experience of domestic household work in describing it as 'just an ongoing chore ... you know ... like washing your face, that just gets done'.

Asked specific questions about domestic labour tasks, women almost invariably recounted the tasks performed using the third person, and passive rather than active voice, thereby obscuring their active engagement in domestic household labour so that the labour tasks appear to perform themselves, or just happen. While women's primary responsibility for domestic household labour is a shared feature of their working lives, whatever other labour activities they do, experientially it becomes a facet of everyday existence that is part of being a woman rather than an identifiable job of work. This 'naturalisation' of women's responsibility for domestic work is central to the ideological process of legitimation. All the women I talked with *hated* domestic household tasks, particularly housework and laundry, which they felt to be boring, unstimulating and unrewarding across the whole spectrum of farm types. None the less they accepted as 'facts of life' or 'the natural order of things' that these tasks were their responsibility and that their husbands did not share them. For example:

Gayle Brown

> I hate the housework and always have. I have to do that myself as no one else does it. I think perhaps that's why I've got so many jobs [in the farm business] because I don't like doing that . . . I only do them because no one else does and then the whole thing falls down.

Julie Church

> There's a conflict of roles all round . . . because the cows I'm supposed to be in charge of and as I say I should be out there every day looking at them and checking them but I'm not because I'm doing other things. I'm either doing the flowers [she has a dried-flower business] or, like this morning, I just did housework which is very boring, I really do hate that, but one has to do it.

Jill Watson

> I'm better if I'm pushed, I'm terrible for being .. I mean I don't like housework, the more time I've got the less I do. I've always worked best under pressure. I hate housework anyway, I mean the more you do the more it gets mucked up anyway . . . I don't feel I could justify having somebody to help out [with the housework] just teaching [she is a supply teacher] because all that money's gone really, but if I can make enough on babygear [her second-hand baby equipment business] to pay somebody . . . I wouldn't mind that.

A low value is attached to these tasks in terms of skill and interest by women themselves as well as socially, in the family household as a whole. This is an important part of the process by which domestic labour is rendered invisible in terms of everyday concepts and perceptions of work. As Julie Church put it, 'It's just taken for

granted that it's your job anyway and that doesn't count [as work] . . .
it's just one of those things that you take on with a husband.'

However these very negative experiences and attitudes towards
domestic tasks relate to a simultaneously strong identification with
the 'domestic roles' of being 'a good wife and mother'. These roles
were a major source of self-esteem. While there is a much greater
diversity of experience between women in different farm types with
respect to these roles, certain elements are shared. Most important, a
sense of 'having to' do domestic household tasks is a powerful one in
which women's experience is founded upon the social bonds of duty
and obligation structured through kinship practices and the emotional
dependencies built up within personal relations. It is through these
obligations and dependencies to particular individuals that patriarchal power relations are translated into women's everyday experience as farm wives. A sense of guilt at not performing adequately in
these roles was at least as important in women's acceptance of this
gender division of labour as a positive sense of fulfilment in the
labour tasks involved, even in childcare and food preparation which
were generally held in high esteem by women.

The case study material suggests that the familial gender division of
labour is built upon, and serves to reinforce, a process by which
women's self-identity is bound up in ideologies of 'wifehood' and
'motherhood' which naturalise gender inequalities. However these
ideologies are not fixed or uncontested. The degree to which women
are expected to participate and do participate in the farm labour
process varies in relation to differing familial ideologies and construc-
tions of 'wifehood' within which individual women negotiate their
position on the farm. These variations do not correspond to neatly
bounded ideological 'types'. They do, however, relate to the extent
to which the value systems of household members (and the local
community to which they belong) have been reshaped by the
commoditisation process, along with the relations of production and
reproduction on the farm.

CONTRASTING IDEOLOGIES OF WIFEHOOD

A distinction can be made between the gender ideologies associated
with the labour process in family labour and family business farms. In
the case of the 'family labour farm' the agricultural labour process is
characteristically uncommoditised and is almost entirely based on the

unpaid labour of family members. In this context, farm wives have an established and legitimate working domain beyond the realm of domestic household labour. Here, the 'farm wife' has a distinctive *farm* status. In the case of the 'family business farm' the agricultural labour process is more extensively commoditised, including the labour of other family members, and farm wives here conform more closely to the bourgeois 'housewife' model. Their labour activities beyond the domestic household circuit are a product of incorporation through marriage into their husband's work and lifestyle, of being 'married to the job'.

The extent to which household consumption and production processes are themselves commoditised affects the values attached to domestic labour tasks by women themselves (and other family members). For example, Hannah Green, Sue Price and Gayle Brown, the women on farms with a significant element of subsistence production, and in which they play a major role, held considerable store by producing fresh, wholesome meals for their family every day and derived considerable esteem and satisfaction from this labour activity. Gayle Brown, for instance, despite spending much of her day bulk-cooking meat pies for the farm shop, enjoys cooking the family meal from their own produce:

> I have always enjoyed getting the meal and things because they're all very appreciative and don't mind trying something new . . . I've always enjoyed cooking and they enjoy eating it, so that's no problem.

By contrast, the three women in the more commoditised domestic situations referred to food preparation rarely and without pleasure. For Julie Church, for instance, 'cooking is a bit of a chore as I'd rather be out in the garden doing my own work, so we eat very late, about 8.30–9.00 pm'. Although conflicts arise more frequently between domestic 'duties' and other labour demands in these situations, they are largely resolved by women *without* challenging conventional gender ideologies or the consensus which establishes their responsibility for these tasks. Such a challenge would touch more deep-rooted tensions in the personal relations and emotional ties with which these bonds of obligation are cemented.

Such tensions reflect a fundamental contradiction in women's experience as 'wives' between their place in the 'collective identity' of the family farm and their sense of identity as *individuals*. Significantly, in all cases of conflict arising from women's multiple labour roles, these conflicts are seen to be 'women's problems', to be resolved by women adjusting their workload. Most usually this means giving up, or reducing, those labour activities which compromise their primary duties as wife and mother. A redivision of domestic labour within the overall farm labour process, in such a way that men give up their privileged position with regard to household servicing work, is not considered. The rest of this section looks at some of the ways in which women negotiate their labour conditions within conventional gender divisions and ideologies, and the conditions and consequences of attempts to challenge them directly.

The differences in gender ideologies of 'wifehood' between family farms at different levels of commoditisation suggested above are substantiated by the ways in which women themselves make sense of their experience of the farm labour process. In three of the case studies the consensus surrounding the established gender division of labour, in which domestic household work is seen as women's primary role, remains relatively intact; these are the Greens (Dorset), the Prices (Dorset) and the Browns (MGB). These are the family farms exhibiting the most tight-knit interdependence between farm production and household reproduction processes and in which these processes are least extensively commoditised.

In the other three cases; the Watsons (Dorset), the Evans (MGB) and the Churchs (MGB), this consensus has been strained or even undermined by the commoditisation of the household reproduction process, and/or the farm production process and the value systems associated with them. In these circumstances women's experiences reflect a number of deep-seated contradictions in terms of their identity as individuals and their position as farm wives. These two groups are termed here 'farming women' and 'incorporated wives'.[3] While the domestic political economy of each farm within these two groups varies in detail (see Chapter 7), as does the gender division of labour, they can be distinguished on a number of key points with regard to the gender ideologies within which their status and identity as 'wives' is constructed.

'Farming women'

In the first group, women's labour extends beyond the domestic household circuit to an extensive involvement in the agricultural or related farm labour circuits. This reflects the significance of a subsistence element in the agricultural production process on the farm and of livestock in the production system as a whole, in which women are more extensively involved. There is a territorial dimension to this gender division of labour in which 'gendered work domains' on the farm are defined by the relationship between the farmhouse and the cultivated fields, mediated by the more ambiguous domain of the farmyard. The distinction between the spatial division of men's and women's work experienced by the two groups of farm women centres on the role of the farm kitchen/parlour on the farm.

For farming women the kitchen/parlour is the hub of the farm labour process, the site not only of staple reproductive activities such as food preparation and consumption, but also the reception of farm visitors, the tending of sick animals and the family meeting-place. As Hannah Green, Sue Price and Gayle Brown variously expressed it, 'the house is an extension of the farm'. Hannah Green makes this point the most clearly in explaining why she has not modernised her kitchen facilities:

> You see this is a very old fashioned kitchen . . . where else would you see a sink like that? But you see, if you have a modern sink unit and you need a bucket of hot water for an animal emergency you can't get a bucket into a modern sink.

Sue Price uses her living-room as the farm office and by combining her domestic household responsibilities and agricultural labour in the same site is able better to manage the simultaneous pattern of work she does. This reflects not only the extension of women's work domains beyond the 'domestic' but also the blurring of the traditional division between 'domestic' labour and 'social labour' on family farms where the same labour process can produce valorised or non-valorised products, for household or enterprise consumption, or for sale on commodity markets.

A second distinguishing characteristic of farming women's experience of the farm labour process is that it has much more in common with that of other family members. Here all 'family labour' is experienced as an habitual practice ('like washing your face') and the labour process is characterised by a highly routinised schedule of daily activity within a seasonal pattern. In this sense, while women enter the farm labour process on unequal terms, structured by patriarchy, their everyday labour experience is much less sharply divided from that of other family members. This is in marked contrast to the experience of 'incorporated wives', where the integrity of the family labour process has been undermined by the commoditisation of farm labour relations.

Farming women negotiate their labour participation within a consensual framework in which their responsibility for domestic household labour is accepted, but as only a part of their position in the 'collective identity' of the family farm. There are two aspects to this. First, the status of domestic household labour, such as food preparation, is itself higher than in farm households with a more highly commoditised reproduction process. The ethos, or value system of the household is centred on the domestic or family *labour* process in which each member's contribution is weighed as part of a collective whole. Food production for household consumption, as part of the subsistence circuit of the agricultural product of the family labour process, is no less valued than other labour tasks. Second, in this context women have adopted a labour strategy or rationale within which they negotiate their daily labour activities in such a way as to *limit* their workload to a series of preferred task areas – those in which they have the greatest freedom of manoeuvre and self-determination. Some sense of this is conveyed in the following extracts from conversations with the women in these three households.

Hannah Green

[in a conversation about silage-making which was currently under way on the farm] I don't get involved now if I can help it . . . When you've done it all your life, and I've worked in the farming community all my life, it's nothing new . . . you have better ways of using your time.

Sue Price

I used to help with the milking before I left to go to work [when she taught at a local agricultural college] when we were very short of labour . . . I've purposely made sure I don't know how to use the new parlour . . . because I think you can make yourself too useful.

Gayle Brown

I leave that side of it to the men . . . the actual farming part, because . . . well I used to do it but . . . I've mucked out at 5am, I mean I worked with Brian [husband] right the way through, but . . . the more jobs that you do the more you're expected to do . . . not in a nasty way, just because you are another pair of hands. I'm not that interested in that side of it, I've not had the experience that he's had . . . it's much better if you do something that you know about and you know my family are all butchers and I know more about that . . .

Significantly, in all three cases in this group, women are not only familiar with the agricultural labour process on the farm but have the skills and experience to do the task which their husbands, or other family members, carry out. Hannah Green, for example, ran the farm for two years when her husband was incapacitated by an injury. This is in direct contrast to the strong sense of exclusion and ignorance which women on the more commoditised farms feel about the agricultural labour process, accentuating their material dependence upon their husbands. Whereas 'farming women' are keen to reduce their agricultural workload or at least to concentrate on those aspects of the farm labour process which interest them most, 'incorporated wives' are keen to learn all the agricultural skills they can to increase their involvement on the farm and/or to break out of the limiting domestic roles to which they feel restricted.

'Incorporated wives'

The second group of women, identified above as 'incorporated wives', live and work on farms characterised by more highly commoditised household and/or farm production processes. In these situations, women are much more clearly incorporated into the farm labour process by virtue of marriage, in the sense of having their working lives structured by their husbands' occupation. In the case of farming this process of 'incorporation' is heightened because, as a number of women commented, not only are they married to their husband's job but, since there is no geographical separation between home and work in farming, they are also permanently 'living on the job'. Again, the individual situations are very different, but the ways in which the conflicts arising from women's more restricted domain of 'legitimate' labour activity are resolved share some common features.

The first common feature is that the territorial dimension of the gender division of farm labour is more sharply defined and inscribed in the physical structure and visual appearance of the farm. Gravel drives and tended gardens mark off the 'domestic' domain from the 'working farm', reflecting a more rigid gender division of labour between domestic household and other circuits of farm labour. In these cases the farm kitchen/parlour marks the boundary between the principal sites of domestic household production and consumption and the agricultural and other commodity production processes. In Julie Church's experience, for example,

> Farming is a way of life. Men like to be on the farm even when there is no work to do. They don't like to be at home. Arable farmers definitely go out all the time, my husband is hardly ever at home.

Moreover she experienced this spatial dimension of the gender division of labour on the farm in terms of a reinforcement of her domestic identity and role and a lack of any 'personal space'.

> It used to be difficult, very difficult, especially when the children were young, because I used to get very frustrated with literally having to mind them the whole time. I just wanted a little while to go shopping on my own. Just to ... do anything on my own and I used to find it very trying that he didn't get home until long after they were in bed. I can't see my husband changing ... I can't see him being at home more because he's not the type.

The second feature common to this group of women's experience as farm wives is the breakdown in the consensus surrounding established gender divisions and labour roles. For example, when Vicky Evans took over the agricultural labour tasks on the farm during the winter months of 1986, while her husband earned some much-needed off-farm income to see the household through a 'bad patch', she began to reassess the division of labour between them:

> I'm not quite sure whether it is working out right or not. I mean I'm running the farm *and* doing everything that I was before ... see I'm not ... I don't think it's quite fair. I don't know whether it's just me being biased. He's working up there [the garage] so he's doing less on the farm, so I'm doing more on the farm so I should be doing less in the house ... I don't seem very good at working out a strategy for sharing more of it [domestic household work]. I can see that it doesn't seem fair to me but he doesn't seem to realise it. I think to Richard children and the house are the woman's job and that's it, even though I'm doing his job as well. I tend to find that he doesn't really understand, so he gets cross and I just get fed up ... he doesn't shout but he seems to think that it is unreasonable for me to discuss it. I've tried occasionally, but it's

not really worth it. I must admit I wonder what he thinks when he sees at least a couple of husbands of friends out in the kitchen, mucking in.

The catalyst to this reassessment was the change in material circumstances on the farm, but it brought to the fore doubts which Vicky Evans had raised, far more tentatively, during earlier visits. It was specifically the contrast with a different set of values and practices with respect to gender identities and divisions of labour in the households of friends, which were taken to represent a more modern approach, which made her query her own experience. Julie Church was likewise frustrated by having to accept primary responsibility for domestic household labour and referred again to non-farming friends who had a more equitable household division of domestic labour:

> What aggravates me so, is that just when I get into it [her dried-flower business] I have to go to get the lunch or go and get the children . . . do something else . . . because when you haven't got lunch or supper . . . I came in last night it was . . . 9.00pm by the time I got in and I just didn't feel like getting supper at all but everyone came in at 9.30 and 'where was supper' and 'why wasn't it ready'. Even Laura [her daughter] was sitting there starving and I said well why don't you come and help me, but oh no . . . It's difficult to make a farmer take the same role as perhaps a man and a woman working in the City because they both come home on the train at roughly the same time and somehow they would have a certain togetherness that they'd get supper together or take it in turn. I have cousins who actually do that . . . but here, if I turned round and said get your own supper it's your turn, I don't know what would happen . . . it just wouldn't work because . . . even early on . . . he'd [her husband] just have turned round and said well all right then, give up the flowers then and get the supper instead. I should be doing that in my spare time and it's got too big now . . . and what's gone is the garden and the cleaning . . . that's whats suffered.

Both Vicky Evans and Julie Church also make clear that, while at one level the onus of responsibility for domestic household labour on women is consensual, at another it is explicitly enforced when that consensus breaks down or is challenged. In a very material sense women's position on the farm and in 'the family' is founded upon an acceptance of profoundly unequal gender identities and inequitable terms of entry into the labour process. Any breakdown or challenge in these gender relations *potentially* threatens the marital relationship and the emotional ties which that entails, and, simultaneously, the viability of the farm as a family enterprise. Thus the explicit enforcement of traditional roles extends beyond the conjugal household and to the kin of the husband, where they share an interest in the property and business assets of the farm and its passage down the family line.

This is clearly illustrated by the experience of Julie Church, whose husband's family retain a strong interest in the farm business. They place considerable pressure on her to restrict her working at her flower business when this is seen to conflict with her servicing tasks and identity as wife and mother.

> I set it [the flower business] up originally well . . . to go on holiday with say, and now . . . I've now got to the stage where even though I've made a profit to go on holiday, I can't take it out because if I do I've got nothing to spend . . . to pay for this year's seed and fertilizer and everything and of course that's what happens on the farm. If you take it out you have to borrow, which costs you then, so you leave it in to save you borrowing, so you're not going on holiday, so you've defeated the object of taking it on in the first place. That's why they [her husband and in-laws] say it is worth my doing it. Seeing as I enjoy doing it I think it is.
>
> It's a good thing but maybe I'm not a very good organiser . . . so I'm still trying to do too much in a way, it's certainly made life more complicated and I was thinking today there's so many things I ought to be doing in the house . . . curtains here or cushions there or something should be done and I don't do it now because I'd rather be out growing . . . They [her husband and in-laws] query whether I should be doing it at all and why I should *want* to do anything like that . . .

Here the spatial division of labour can be seen not simply to reflect the gender division of labour on the farm but to be a constituent of it. Men's authority over space, deriving from the patriarchal property relations examined in the last chapter, underlies one practice by which men actively control women's entry into non-domestic labour circuits and reinforce their dependent status, thereby enforcing patriarchal labour relations. In Julie Church's case, her husband and father-in-law restrict her ability to develop her dried-flower business on a fully commercial basis, whether intentionally or unintentionally, by refusing permission to move the 'making-up' process and administration of the flower business from cramped space in the attic to a large ground-floor room in their married home, which is currently one of three under-used farm office.

Vicky Evans and Jill Watson, while at opposite ends of the commoditisation spectrum in terms of the agricultural production process, face similar contradictions between their labour practices and gender identities as wives, arising from the dependence of the household livelihood on money-income from off-farm wage labour. In Jill Watson's case, where the agricultural production process is now marginal in terms of supporting the household, it is her income from supply teaching which primarily meets the household's consumption requirements and repays the farm's debts. She enjoys this work at one level because 'You tend to stagnate just with children, and need something to take yourself beyond that . . . like teaching, when your mind's alive all the time.' But at another level it presents difficulties in terms of dealing with her husband's insecurities concerning this 'role reversal'.

In Vicky Evans's case, where the agricultural labour process is itself highly integrated into the wider money economy, there has not been such a dramatic 'structural' reversal of roles in that it is still her husband who earns and controls the household's money-income. However, in terms of her experience of the farm labour process, it has undermined her own justification for the inequitable division of domestic household labour and, to a lesser extent, her acceptance of this as women's 'proper role'.

By exploring the implications of these contradictions in the future tense Julie Church is able to express a very difficult personal dilemma arising from the subordinate nature of the gender identity of wifehood. She describes her life as guided by the principle that 'I drop my own interests if the family needs me, the family comes first . . . I've got to have that principle I suppose.' It is a principle which defines her only in relation to others, producing an identity and existence which she recognises as, at best, precarious:

Although he's [her husband] not here, the children are taking his place, which means that when they leave home I'll be more lonely because he's got into the way of going out and the children will leave anyway so I'll be on my own. All your good years you're so much in demand you can't do all the things they want you to do,

and when you're worn out, presumably, you're dumped [laughter]
. . . I reckon I've got to make my own life, which is one reason why
I think I do the flowers, because if I don't make it, I'm going to be
left sitting at home here either worrying about him or I suppose
getting neurotic or drink or something. There's loads of things I
want to do anyway. Actually I think that he'd be quite happy if I
went round with him, but I don't think that I would be 'cos I don't
like just sitting not making any decisions all the time. He takes the
dog with him now. Well I don't want to be just a replacement for
the dog [laughter].

PROCESSES AND PROBLEMS

The analysis of gender ideology in this chapter has provided some
answers to questions raised previously about the legitimisation of
patriarchal labour relations and practices in family farming. It
suggests a complex process by which ideas and meanings about
appropriate gender divisions and roles are constructed and contested
in the labour practices of family farms so that women as well as men
are implicated in this process, and gender inequalities are legitimised
by consensus as well as coercion. However the analysis also raises a
number of difficulties in trying to deal with the complexity of the
processes examined.

The analysis shows how women themselves represent their work in
ways which commonly undervalue it or discount whole aspects of it as
somehow not 'proper' work. It has traced the way in which women's
experience and practice of doing 'several jobs at once' in part
influences this tendency. However the implication is not that women
are somehow 'duped' into acquiescing in their own subordination but
that the relationship between the ideas and meanings constructed
through subjective experience of the labour process, and the ways in
which these ideas and meanings inform labour practices themselves,
is considerably more complex than notions of 'bias' or 'false consciousness' allow.

The interpretation of ideology arising from the analysis of the case
study material is of a process of everyday sense-making in which the
reference points informing both women's accommodation and contestation of gender inequalities in the family labour process are
constructed from within a known lifeworld, or what might be called
concrete knowledge, as opposed to abstract knowledge in the shape

of an objective/rational representation of social reality.[4] The analysis highlights three features of the construction and mobilisation of patriarchal gender ideologies in the family labour process which legitimise women's subordination as farm wives.

Firstly, it has identified a complex web of meanings in which domestic household labour tasks and relations are defined as 'women's work'. Ideologies of wifehood are central to this web and to the process of 'naturalisation' which legitimises patriarchal labour relations by redefining domestic household labour, not as work, but as a role intrinsic to women's gender identity. It is suggested that it is less 'domestic' tasks as such which define what is distinctive to women's work experience, as the meaning of these tasks can change, depending on the labour relations under which they are carried out. Rather, it is the labour relations specific to wifehood, which build upon and reinforce women's subordinate status, which are distinctive.

However, in exploring these questions, the analysis calls into question the assumptions underlying the initial contention that ideologies of wifehood which legitimise gender inequalities in the family labour process are constructed solely within the labour process itself. The subordinate gender identities which are mobilised in the farm labour process are seen to be constructed through a range of practices, such as sexuality and emotional dependence, although the case study material deals only tangentially with these issues.. The implications of this for Burawoy's wider argument, siting the construction of compliant working-class ideologies in the labour process itself, are discussed in the general conclusions.

A second conclusion to be drawn from the analysis is that gender ideologies of wifehood are not fixed. Two ideologies of wifehood have been distinguished on the basis of differences thrown up by women's own articulation of their experiences of the farm labour process.[5] The terms 'farming women' and 'incorporated wives' used to define these two groups indicate an underlying shift in women's experience, from a central position in a household production system centred on family labour and organised around the conjugal household, to a marginal position in a household production system centred on family capital and organised around patrilineal kinship relations. This suggests that the idea of *a* patriarchal gender ideology (for example, Delphy, 1984) is misplaced and that the normatively approved terms within which women work need to be analysed historically (see Seccombe, 1986).

Thirdly, the analysis of the case study material distinguishes between two related processes underlying the transformation of gender ideologies, a dialectical relationship between gender practices and ideologies themselves, involving direct contestation between men and women, and the commoditisation of the relations of production and reproduction of family farms and of the value systems of farm families. However, in focusing on the farm labour process and family household, the case studies do no more than raise questions about the significance of 'local' value systems and gender ideologies for the way in which these processes operate within the family farm. How is it, for instance, that Vicky Evans and Julie Church voice their discontents in terms of 'modern' (urban) values and practices while Hannah Green and Sue Price openly reject such 'norms'? Within the gender regime of the family farm, what the case study analysis does demonstrate is that any renegotiation of the gender division of labour takes place within structurally unequal power relations between men and women, which include men's enforcement of patriarchal labour practices and relations.

Patriarchal gender ideologies, it has been argued here, are woven into the fabric of the family enterprise and the family labour process and are material to the perpetuation of women's subordination within them. In this the analysis supports wider arguments (see, for example, Cockburn, 1986; Davidoff and Hall, 1987) that material practices and social relations not only gain expression in ideological terms, but that ideology bears upon and influences material practices. It is not suggested that the relationship between gender ideologies, patriarchal labour relations and the commoditisation process is neatly synchronised, but that each has a different rhythm such that they interrelate in sometimes contradictory ways. The next chapter turns to examine the nature of this relationship in practice within the wider political economy of six family farms, characterised by different levels of commoditisation.

7 The Domestic Political Economy of Six Family Farms

The case studies presented in this chapter develop the themes raised in the previous analysis by examining some of the ways in which patriarchal labour relations are taken up, built upon and reshaped by the commoditisation process. The analysis examines the domestic political economy of six case study farms in terms of the framework set out at the end of Chapter 3. This includes the household structure of those living and working on the farm; the composition of the principal and subsidiary labour circuits on the farm in terms of the participation of family and hired labour; linkages with external capitals in the farm production and reproduction processes; the circulation of income and money-capital between household and enterprise and the structure of capital and land ownership on the farm in terms of kin and, where appropriate, non-kin interests.

The analysis highlights the varied relationship between the processes of production and reproduction in the family household and the agricultural enterprise, at different levels of commoditisation. It focuses on the intersection of class and gender in shaping the internal relations of the farm and seeks to demonstrate the reflexive and sometimes contradictory relationship between patriarchy and commoditisation. Familial gender divisions and labour relations are shown not only to be restructured in the commoditisation process but to be active constituents of the commoditisation process itself. While centring on the conjugal household at the core of the family farm, the analysis situates the domestic political economy of each case study within a wider network of social and economic relations in which it is embedded. The aliases used in the previous chapter are retained to preserve the anonymity of those involved.

The analysis draws on material from all the components of the research methodology set out in Table 4.1. The case studies are organised to exemplify a range of different configurations of patriarchal and commodity relations in the farm labour process across the typology matrix. The first section presents a case in each of the categories of 'family labour farm' (category A) and 'family business farm' (category C). The second section looks at two 'transitional' (category B) farms, while the third re-examines a case from either extreme of the spectrum of commoditisation, where the integrity of the household production system is breaking down as household and/ or agricultural enterprise are integrated into the wage-labour economy. Each section is preceded by a short introduction identifying the main themes guiding the analysis of each of the paired case studies.[1]

This presentational strategy is designed to highlight the active role of human agency in the commoditisation process and the diversity of patriarchal practices associated with the farm labour process within, as well as between, different structural contexts. The case studies are illustrative of themes addressed earlier and it is not intended to draw any general conclusions from them. However the analysis does raise a series of more general analytical and political questions about the relationship between the personal experiences and individual practices examined and the construction and transformation of patriarchal gender relations as social structure. These issues are assessed in the concluding chapter in terms of the contribution of the case studies to the wider theoretical and methodological arguments put forward in the book.

TWO FACES OF FAMILY FARMING

The first two cases represent examples of a family labour farm and a family business farm exhibiting very different levels of commoditisation in their relations of production and reproduction and in their associated value systems. The gender divisions of labour, property rights and money-income are markedly different between the two cases. Mrs Green and Mrs Church work within contrasting ideologies of 'wifehood' associated with the two categories of 'farming women' and 'incorporated wives' identified in the previous chapter. Comparing these two cases draws out the significance of the shift from labour to capital as the organising principle underlying the structure of domestic commodity production and the shape of patriarchal gender relations. Where the Greens demonstrate an apparently cohesive unity between household and enterprise, patriarchy and commodity production, at both the level of personal and social relations, the Churches demonstrate a range of tensions between them, likewise manifested at both personal and social levels.

The Greens at Holly Farm: a family labour farm (type A/A)

The Green family have lived at Holly Farm in west Dorset since 1956 when Mr Green bought the core holding as a going concern for £8500, incorporating 50 acres, farmhouse and dairy yard buildings, 22 Ayrshire dairy cows and 20 followers,[2] two tractors and working implements. In 1962 he bought a further 26 acres of off-lying land for £4800 in order to build up a pedigree Jersey herd and expand production. In the mid-1960s he invested in a modern milking parlour, using money from the sale of a cottage on the off-lying land and of half an acre of roadside land for housing development. Since then he has also replaced one of the tractors and put up a new silage tank. Between 1974 and 1983 he rented a further 45 acres of land from an adjacent farm. This represented the farm's peak in terms of its productive capacity at 121 acres, with a dairy herd of 30 cows plus followers. It coincides with the period when Mr Green's son was working on the farm full-time and the expansion was preparatory to his taking a full stake in the holding and to its supporting two households until Mr Green retired.

In 1983 this prospect evaporated when a family dispute took place between Mr Green and his son and daughter-in-law over setting up a one-third/two-thirds partnership in the farm. The offer was eventually presented as an ultimatum to the son (whose wife, Mrs Green and her daughter allege, was pushing him to hold out for a 100 per cent stake and to 'get it in writing'). When his son did not respond, Mr Green sold half of the stock and gave up the tenancy on the 45 acres as a symbolic means of revoking his son's claims to the farm. As their daughter put it: 'The sale was, in my father's eyes, as much as to say that's definite . . . and he had to go through with the sale to prove to them that that was it.'

The son is no longer involved in the holding but is now on speaking terms again with his parents, the burden of responsibility for the episode having been shifted by the Greens onto their daughter-in-law. They regard her as an 'outsider' because she is not from 'farming stock' and does not hold what the Greens call 'farming values' – defined as being 'family orientated' rather than 'money orientated'. These metaphors distinguish her as a non-blood relative and are overlain by the fact that her behaviour challenges the 'natural' gender order of marriage which, in turn, reflects badly on their son. As Mrs Green explains the episode, 'He's idle and his wife is very domineering, so he didn't stand up for himself.'

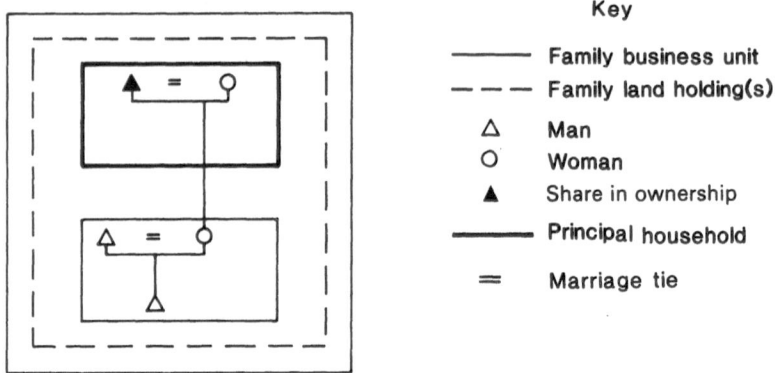

Figure 7.1   Household structure at Holly Farm

The current political economy of the farm is centred on the family labour process which is embedded in generalised commodity circulation for the realisation of the value of what it produces, but is relatively self-contained in reproductive terms, with limited commoditisation of household consumption and agricultural production. The family labour process is principally structured around Mr and Mrs Green, now in their 60s, who live on the farm but also incorporates the regular labour of their youngest daughter (Jane) and her husband, who are in their early 20s and live a quarter of a mile away in the village with their infant son. Figure 7.1 shows the household structure of the farm and indicates that, whilst two households and four individuals are involved in the farm labour process, it is Mr Green alone who owns the business capital with which they work. He is also the sole owner of the land assets (76 acres in two pieces), except for 7 acres of pasture which is attached to his daughter and son-in-law's cottage.

The farm labour process comprises only the basic circuits of domestic household production and agricultural production. The main agricultural enterprises are dairying, with a herd of 15 Jersey cows (and two house cows to meet the household's own milk requirements), a beef sideline from the dairy operation and a flock of 50 ewes producing lambs for meat. Chickens, turkeys and eggs are also produced on a small scale, principally for household consumption. Associated with these livestock enterprises, both hay and silage are important seasonal activities, and potatoes are grown as a market crop on some 10 acres of land. As Figure 7.2 shows, the farm labour

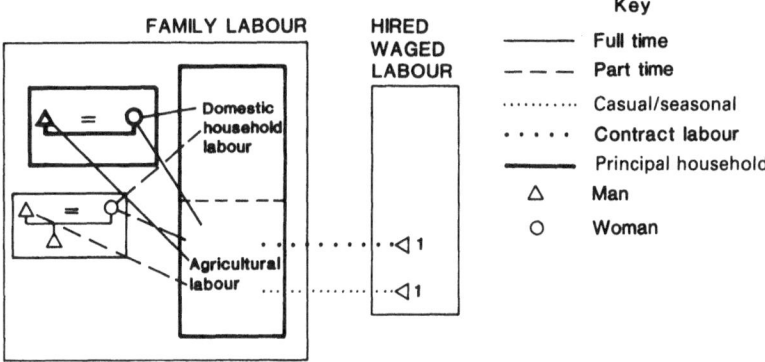

Figure 7.2   The labour process at Holly Farm

process operates entirely on family labour, employing only one additional hired worker at haymaking and contract labour for the specific task of hay baling, as the farm does not have its own baler.[3] There is, however, an established network of co-operation between the Greens and their neighbours, a larger farming concern, at seasonal peaks such as hay- and silage-making and the potato harvest.

There is a marked familial gender division of labour both between the two labour circuits and within the agricultural labour process. Mrs Green does domestic household and agricultural labour tasks in tandem; Mr Green does solely agricultural work. A similar pattern of work characterises the contribution of their daughter and her husband, but on a part-time basis. However Jane does a range of agricultural tasks, such as the afternoon milking and helping her father in the husbandry operations in the fields, which her mother rarely does, and is treated in this respect as a surrogate for her brother. Unlike him, however, she also sometimes relieves her mother in preparing meals for the family members working on the farm.

Associated with these broad gender divisions in labour tasks are inequalities in the conditions of labour participation between men and women, structured by the conjugal and filial ties between them. For example, despite being a trained dairy maid, Mrs Green never does the milking except in an emergency. The farm jobs she does are either 'servicing' this main function, like bringing the cows into the milking parlour and cleaning out the sheds, or else in less commoditised sectors of agricultural production, such as poultry and eggs,

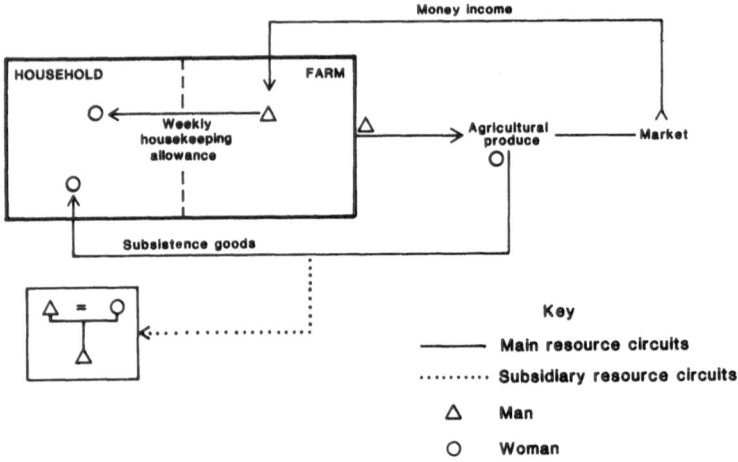

Figure 7.3　Budget structure at Holly Farm

which are produced mainly for the family's own consumption and only casually for local sale. This means that Mrs Green is marginal to the valorised sectors of production on the farm and to the main technological means of production associated with it. In consequence, she is also marginal in terms of the income-generating activities of the farm. In contrast, all Mr Green's labour is in the valorised sectors of production, except for the vegetable garden which he keeps as a hobby and which contributes to the household's subsistence budget. Figure 7.3 shows the circulation of money-capital and income between household and enterprise expenditures, indicating the central position of Mr Green in mediating these flows.

A single pool of money income services both household consumption and agricultural reproduction expenses. There is no financial separation in the form of a wage, salary or fixed payment from the farm account to a household or individual account. The farm account *is* Mr Green's account. The overall significance of money income to household consumption, however, is remarkably low, with a high level of self-sufficiency in terms of food and a life-style based on very low levels of conspicuous consumption. Mrs Green relates this self-sufficiency with pride:

> We grow all our own vegetables, eat our own meat [butchered locally], freeze our own fruit . . . we have our own milk and eggs and I make jam and bread and cakes . . . we even mill our own

flour sometimes . . . so it's only really sugar and coffee and things that we buy in.

By and large, Mrs Green has retained traditional food production methods and the purchase of consumer durables is limited to technological aids in the food production process, comprising two freezers for the storage and preservation of seasonal foods grown on the farm. More recently, a new washing machine was purchased by Mr Green for the household. The purchase of such equipment is weighed against agricultural requirements and drawn against the same account. However, unlike agricultural items, purchases of such household equipment are regarded as a gift from Mr Green to his wife. As Jane put it, 'Daddy treated Mummy to a new washing machine.' Mr Green's control of the household's links with the money-economy is supplemented by his having an independent source of income (estimated at about 20 per cent of total farm income) from share investments. Mrs Green's financial dependence on her husband is reinforced by her lack of direct access to money-income, her only 'independent' income coming from her 'leisure' interests in horses, which she occasionally buys and sells, reinvesting the proceeds in their upkeep.

The purchase of regular necessities for household consumption is managed by Mrs Green from a weekly housekeeping allowance provided by her husband from the farm account. Her control over the domestic economy is increased, however, by her direct access to the non-commoditised circuits of value on the farm through her primary role in the subsistence sectors of the agricultural labour process and the food production tasks associated with domestic household labour. Both Jane and her husband earn their living primarily from waged employment. She supervises school dinners and he works at a local mill on night shifts. Their main link into the farm budget is through the circulation of subsistence goods, principally food, both through meals shared in her parent's household and produce given to them for their own freezer.

The Churchs at Naylors Farm: a family business farm (type C/C)

The Church family have farmed in south-east Essex since 1924 when the current occupant's grandfather bought an initial 150 acres of land at Colston. Over three generations the family have built up a sizeable business consisting of nine farms organised managerially in two holdings and operated financially as one business. In 1932, the original holding was expanded by renting 500 acres from an institutional land-owner at Barstock. In the 1950s it was further expanded by the purchase of farms between their Colston and Barstock holdings as they came onto the market, reaching a peak in 1970 of 1510 acres. They now farm some 1400 acres, all of which, except for Barstock, is owned in some form by the family. Underlying this rapid expansion is a strong entrepreneurial ethos which pervades the family's farming rationale and their value system more widely. Land is seen as a flexible asset and farming, although predominant, is seen as only one of several potential uses. The pattern of expansion and development characterising the family enterprise is based on a land management strategy which takes full advantage of the opportunities for investment and profit arising from their urban-fringe location.

Three households, spanning four generations of the Church family, currently live on the farm and hold an interest in the land and business, although only Mr Church and his family household, living at Naylors Farm, derive a living from agriculture directly. The case study focuses on this household, which consists of Mr Church and his wife, who are in their early 40s, and their two teenage children. Figure 7.4 shows the current household structure of the farm.

Despite their close proximity, the social contact between these three households is conducted largely through the property interests in the farm business and land assets which link four members of the Church family spread between the households. This link takes the form of a partnership set up in 1970 between Mr Church, his father,

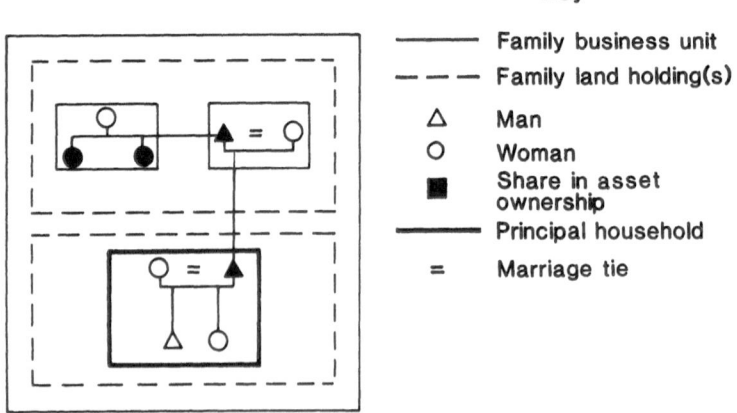

Key

——————	Family business unit
— — —	Family land holding(s)
△	Man
○	Woman
■	Share in asset ownership
——————	Principal household
=	Marriage tie

Figure 7.4   Household structure at Naylors Farm

mother and aunt. The partnership was established when his father was still a young man as a 'tax efficiency' measure and a means of securing inter-generational control of the farm assets from father to son. Managerial control of the farm was transferred to Mr Church from his father, along with the tenancy and residence of Naylors Farm. Mr Church senior has been chiefly involved with the off-farm enterprises, although he still plays a key role in the financial management of the farm. Mr Church's aunt, who lives with her mother and sister in another of the houses on the farm, does all the business accounts for the partnership. His mother is a 'sleeping' partner in the business, with no formal or active role in its affairs. The partnership owns 350 of the 900 acres of in-hand land on the holding. The rest is split between different individual family members who nominally 'rent' it to the farm business.

The current economic structure of the farm is highly complex. What began as a mixed agricultural holding with a large dairy unit is now almost entirely arable, with an increasing diversity of non-agricultural enterprises off the farm, such as a garage and a calor gas business on land they used to farm at Denwood. Attention is focused here on the agricultural labour, domestic household labour and non-agricultural farm-based labour circuits. All three circuits are highly commoditised. Both household consumption and agricultural production are extensively tied into the wider market economy and dependent upon external circuits of capital.

Agricultural production consists of large-scale arable cultivation, principally rape, wheat, beans and peas, and barley (in order, by value) and an intensive beef unit. The most important non-agricultural farm enterprises are herb growing and the bulk manufacture of mint and horse-radish sauces, which are run by Mr Church, and flower growing for dried-flower production run by Mrs Church. In addition, income is generated from the renting out of land and buildings for numerous non-agricultural uses which directly subsidise the agricultural and flower enterprise accounts (see below). Production is highly commoditised in terms of its dependence upon technological inputs and 'scientific' labour from agrotechnology manufacturers. Mr Church is a member of a large Essex farmers' co-operative which uses its collective buying power to improve its market position, cutting out the merchanting sector of the inputs chain. In addition, much of his production, specifically beans, peas, mint and horse-radish, is grown on contract for food processing and packaging companies.

Farm labour relations are, likewise, highly commoditised with

14 full-time hired workers, three part-timers and about 20 seasonal/ casual workers.[4] Mr Church works full-time on the farm at the agricultural and herb enterprises, largely in a managerial capacity. Their son, now aged 13, has also started to assist his father on the farm and is keen to take on the business when he is older. Apart from two tractor drivers, all of this labour, whilst principally employed in the agricultural production process, is used flexibly between the non-agricultural enterprises as they can be spared from agrulcultural tasks. However it is Mr Church who takes the labour deployment decisions. The enterprises which he manages, the agricultural and herb enterprises, take priority, followed by his father's labour demands for his non-agricultural enterprises. Mrs Church's dried-flower enterprise, which is dependent on seasonal hired labour, takes lowest priority, competing with the deployment of farm labour in the maintenance of the buildings and grounds of the house. Figure 7.5 shows the labour structure of the farm. The off-farm sectors of the business are too complex to incorporate in detail, but are indicated in terms of the deployment of farm labour.

Mrs Church is almost entirely excluded from the agricultural production process, incorporated only by default through dealing with personal callers at the house and telephone messages connected with the farm business which she relays to her husband on a Citizens' Band (CB) radio. Her primary labour role is in the domestic household circuit, providing the daily servicing associated with family reproduction. However the house and grounds are more than the site of household consumption and play an important role in maintaining the status of the Church family in the local community. Mrs Church's

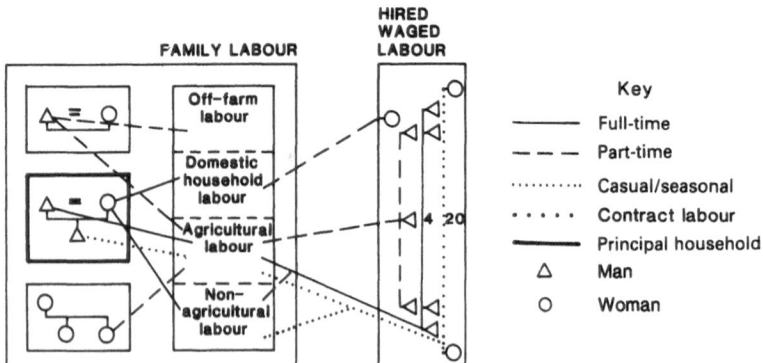

Figure 7.5   The labour process at Naylors Farm

'domestic' responsibilities thus extend beyond the reproduction of household labour to the reproduction of her husband's local status and 'family name' through the regular extension of her domestic skills, particularly food preparation, to his business and shooting friends, her children's swimming and skating parties and charity events. For example, she describes

> two cricket matches last week that we did the food and barbecue for ... they wanted salads and cakes and pies and things ... so I didn't do anything on the flowers last week. If he wants me to do shooting dinners in the winter that's another big ... there's so much work. There's a syndicate of about nine or ten farmers ... it's just a big cooking effort, that's all it is, they're just gluttons. It takes a day to prepare it, and a day to cook it and do it all, and virtually a day to clear up.

Domestic household labour on the farm has also been commoditised to some extent – but only those tasks related to the maintenance of property. There is a daily 'cleaning woman' who does the 'heavy' housework, while general farm labour is used to maintain the building and grounds. Food preparation, washing and ironing, and tasks associated with her identity as wife and mother, such as taking the children to and from school, remain strictly defined as Mrs Church's responsibility. Her other labour activities include the dried-flower enterprise which she set up as a separate business, using 10 acres of land and two barns on the farm; the administrative work associated with the cattle breeding enterprise, which she learnt from her father; and doing the 'books' for her father's cattle business in Chidwell.

Her conditions of labour are distinct from those of her husband or father-in-law in that there is a major clash between the family's 'entrepreneurial' ethos, which she shares, and the domestic role which she reluctantly accepts as her primary 'duty' and which, as the previous chapter showed, is overtly enforced by her husband and in-laws when it is seen to be threatened. As a result, she runs the flower business on a shoestring, doing all the sales, buying and manufacturing labour in her 'spare time'. She has no direct control over the working capital or labour on which the flower business depends, which her husband and father-in-law determine through their alloca-tion of labour and working capital between enterprises. Moreover the potential for founding an asset or income-base independent from her husband and his family is minimised by his incorporation as a 'sleeping partner' in the ownership of the flower business.

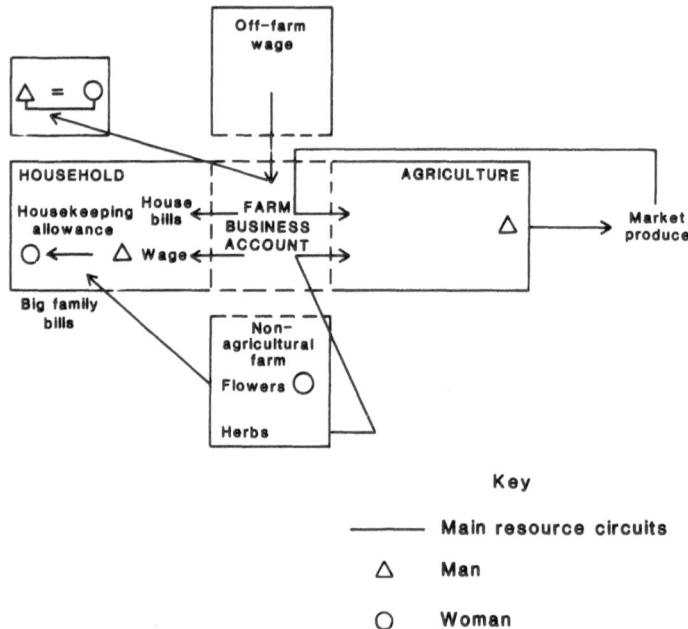

Key

——————— Main resource circuits

△   Man

○   Woman

NB. Subsidiary circuits not marked to reduce complexity

Figure 7.6   Budget structure at Naylors Farm

The circulation of money-income between the various aspects of
the farm business and the different households on the farm is
extremely complex. Figure 7.6 provides a simplified version of it,
focusing on the distribution and control of money-income between
the Church household at Naylors Farm and on-farm enterprises. In
1984/5, the farm business (agricultural and herb enterprises only)
generated a turnover of some £600 000 a year, of which
£400 000 was surplus to outgoings incurred in the production
process itself.[5]

With the exception of the profits from the dried-flower business, all
the household income is circulated via the farm business account,
either directly through the payment of major bills, such as utility
services and house maintenance, or through the salary paid to Mr
Church from the account. From this he provides a housekeeping
allowance to Mrs Church for food and other household consumption
items. She buys all the household's food from a freezer centre, or

local farm shops, except for seasonal vegetables which she grows with
the help of one of the hired farmworkers in slack periods. What is not
given over to housekeeping is kept by Mr Church for his personal
use. Mrs Church only receives a direct money income from her work
in the flower business which she receives in the form of a share in the
profits. Again, this business is itself subsidised by the main agricultural enterprises, as it uses labour, working capital and buildings from
this labour circuit for a nominal payment only.

However, in contrast to the case of her husband, Mrs Church's
earnings from this activity are spent on collective family household
goods, and specifically the £6000 school fees (plus transport
and uniform costs) for the children. It was to cover these costs that
Mrs Church first established the business. Any surplus is spent on
family treats such as skiing holidays. In other words, the income from
the flower business, dependent on Mrs Church's management and
labour, is integrated into the household as a whole rather than 'hers'
to use as she chooses, even if that choice were to include collective
items of expenditure. This is the opposite of her husband's situation,
where the amount from his salary for collective expenditure is fixed
and any collective expenditure over and above that is at his discre-
tion. Mrs Church's material dependence upon her husband is accen-
tuated by the degree of commoditisation of both household and
enterprise.[6]

THE FAMILY FARM IN TRANSITION

In terms of the commoditisation of the internal relations of the farm/
household, the Prices in Dorset and the Browns in the Metropolitan
Green Belt share very similar structural conditions. However, in
comparing these two cases, the analysis highlights differences in the
experience and manifestation of gender and commodity relations
arising from the mediation and modification of the commoditisation
process by the practices of the households involved. The analysis
demonstrates the instability of farms in this category. Unlike the
Greens at Holly Farm, they are too extensively tied into an externally
determined dynamic of accumulation, in terms of both household and
enterprise reproduction processes, to remain static. Moreover the
gender ideologies and divisions exhibited in these family enterprises
are in a state of considerable flux. Both cases provide evidence of the
restructuring of patriarchal labour relations around women's position

as wives in the division of property and in decision-making and of the contradictions arising from this process.

The Prices at Fountain Farm: a transitional farm (type B/B)

The Price family has been living at Fountain farm since 1980 when Mrs Price and her husband entered into a partnership with her father and mother. Mr Brett, Mrs Price's father, had farmed the 300 acres at Fountain Farm since 1950, renting the land from the local estate owned by the Allen family. He had also purchased Moat Farm (50 acres, buildings and farmhouse) in 1967 in order to expand the productive capacity of the farm and to get a 'stake in the land', and this is where he and his wife went to live on the formation of the partnership. The Bretts have a history of farming in the local area going back some 300 years and several branches of the family currently farm in Dorset. Mrs Price was involved in the farm from early childhood as the eldest of seven sisters, but left the farm for some 15 years between 1965 and 1980, when she took an agricultural degree and PhD in dairying at university and later became a lecturer at an agricultural college.

In 1980 she moved back to the farm with her second husband (she is now in her late 30s and he is in his early 30s) when her father, because of his failing health, offered her a partnership in the business and supported her in her bid to take over the tenancy at Fountain Farm. Within three years both her father and mother had died and so she and her husband now farm both holdings on their own. While Mrs Price is in the very unusual position of holding the tenancy at Fountain Farm, she and her husband are joint partners in the ownership of the business. However Mr Brett left all seven sisters a share in the land and business assets at Moat Farm in his will, and Mrs Price is currently negotiating to sell some of the land at Moat Farm in order to use her share of the proceeds to buy out the other sisters' interests in the business. Although all her sisters still live locally, none actively participates in the operation of the farm and only one is involved in farming through marriage. The current household structure of the farm is shown in Figure 7.7 and comprises Mrs Price, her husband and their three young children.

As with the Greens at Holly Farm, the political economy of Fountain Farm is centred on the family labour process, but, while the anti-consumerist values of the Greens are shared by the Prices, the larger scale of their agricultural operation and the pressures of a

Key

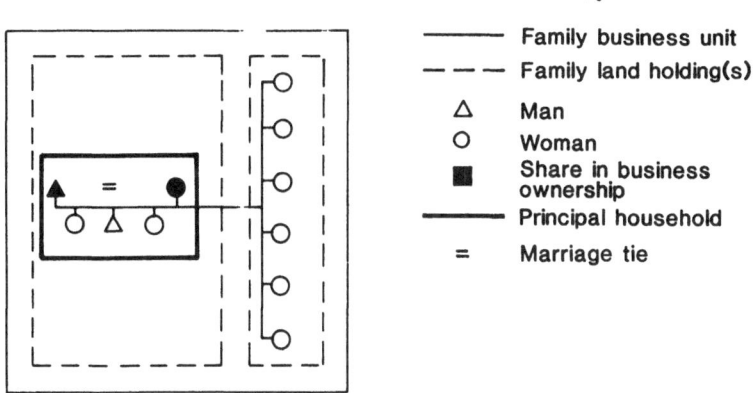

———	Family business unit
— — —	Family land holding(s)
△	Man
○	Woman
■	Share in business ownership
———	Principal household
=	Marriage tie

Figure 7.7   Household structure at Fountain Farm

commercial rent have encouraged the intensification of the production process, so that it is significantly more commoditised and dependent upon external capital in the form of technological inputs and credit. Most significantly, in order for Mrs Price to secure the tenancy, she had to pay 'key money' to the landlord's managing agents in the form of a doubling of the rent. This was initially financed by increasing the level of borrowing from the banks on overdraft and both these external market relations continue to influence the labour strategy and internal relations on the farm.

While the farm labour process remains centred on Mrs Price and her husband, it too is more commoditised than in the case of Holly Farm, partly as a response to the external commercial pressures noted above and partly as a result of the decline in family labour on the holding with the death of her father. Consequently the farm now employs two full-time hired workers and one part-timer, providing about 50 per cent of the agricultural labour on the farm. The main agricultural enterprise is a dairy unit with a herd of about 110 milking cows plus followers. Associated with this is the cultivation of cereals for feed, which is milled for them by a local firm, and of hay and silage for winter feed. The financial position of the farm has been made worse by the imposition of milk quotas, limiting the total size of their production just when they needed to expand to meet their increased financial commitments. Since 1984 this contingency has been met by increasing the quality rather than quantity of milk production. However this in turn has entailed a more intensive regime in terms of feed and supplements and concentrates, raising

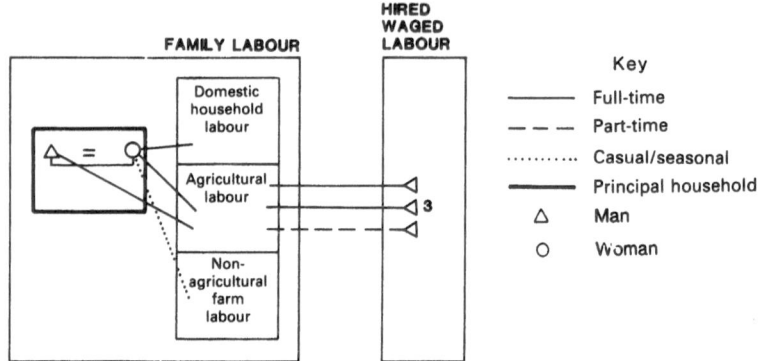

Figure 7.8   The labour process at Fountain Farm

production costs and further compromising the autonomy of the reproduction process. The total turnover of the farm in 1984/5 was about £200 000. Figure 7.8 shows the current structure of the farm labour process.

Once again, there is a marked gender division of labour on the farm between the domestic and agricultural labour processes. Mr Price works solely in the agricultural circuit, both in a managerial capacity and manually on the cultivation side along with one full-time employee, while the full-time 'dairyman' does the milking. Despite her professional training in dairy production and long-established experience at doing most manual agricultural work, Mrs Price is rarely involved in the manual side of the agricultural labour process. Instead, she does all the domestic household labour and runs the business side of the farm, using a computerised accounting and dairy management system and undertaking all dealings with outside agencies. These include solicitors, the Milk Marketing Board, the bank, the accountants and the land agents. This division enables her to combine her domestic responsibilities as a mother with young children and as 'homemaker' with agricultural tasks that are home-based. At the same time, however, this restricts her involvement in outdoor farm work, which she enjoys.

Her position is extremely unusual for a woman. As the analysis of the survey material in Chapter 5 showed, where women inherit land or business assets it is more usual for these to be integrated into the husband's business and it is rare for women to exercise their legal rights over such assets even where they do retain them independently.

This in part reflects Mrs Price's formative experiences of responsibility on the farm, as the eldest child in a family with no male heir and the fact that, as a consequence, she is well known among the local farming and business community as 'her father's daughter'. As she put it:

> I think partly I've benefited because my father hadn't got a son. He was a very unhappy man about not having a son – he accepted it and just brought up each of us to believe we should have a career.

However Mrs Price is very conscious of the distinctiveness of her position and also sees it as part of a continuing struggle against discrimination and prejudice towards women inside farming itself and the agricultural educational establishment. She explains:

> I was the token woman in agricultural college and in the research department and when you've had a lifetime like that and people ask you are you discriminated against I just don't bother to answer.

In the farming context she expands this theme very revealingly:

> I feel it's important in legal terms actually to be the legal tenant. It doesn't mean that Martin [her husband] couldn't do it on his own. It underlies the fact that you do it together I think 'cos otherwise they [banks, merchants etc] do tend to regard the 'little woman' as sort of somebody who sits in the background, and they take advantage. It does mean a rethink among the people you deal with. You still get them on the telephone treating you as if you don't understand a thing and asking to speak with your husband.

Mrs Price not only manages the farm budget in an administrative sense but also controls a sizeable proportion of its distribution in terms of both household and enterprise expenditure. Figure 7.9 shows the budget structure of the farm. Until the arrival of a new dairyman at the end of the research period, Mrs Price supplemented the income from the agricultural business by letting out a farm cottage as a holiday home for six months of the year. This activity is included in the description of the labour and budget structures of the farm in Figures 7.8 and 7.9.

Both Mr and Mrs Price receive their own monthly salary from the farm account, which passes into separate bank accounts. Inequalities do, however, enter into the distribution of their incomes in terms of household expenditure. Mrs Price's salary is the staple source of 'housekeeping' money; that is, it is **spent** principally on commodity

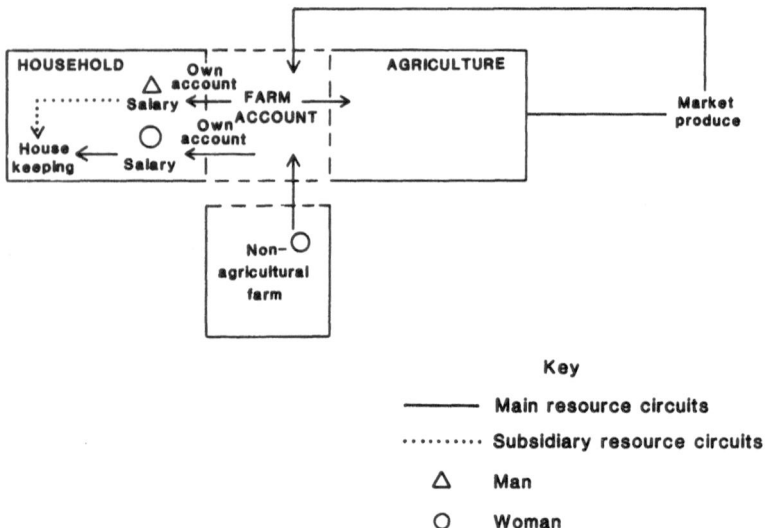

Figure 7.9   Budget structure at Fountain Farm

items for collective household consumption, such as food and children's clothing, whereas Mr Price only devotes 20 per cent of his salary to 'collective' household expenditure and retains the rest for his personal use, such as his field sports hobbies. This inequality is a bone of contention between Mrs and Mr Price – as yet unresolved. Decisions on expenditure from the farm account for major household expenditure, such as items of domestic equipment, and for investments in the agricultural production process, are made jointly. Mrs Price's unusual position in the business and detailed knowledge of its finances are as influential in the farm decision-making process as her husband's practical experience of agricultural work. This is reinforced by the fact that her husband is not local, nor from 'farming stock' and learnt all his agricultural skills at the college at which Mrs Price taught.

If she were a widow, Mrs Price's position would conform more comfortably with normative gender roles in the farming community. It is the presence of a man – her husband – actively involved in the farm, which means that it challenges a range of ideological assumptions and material inequalities in the gender regime of family farming. However a radical interpretation of her position rests uneasily with her own motivations. Her determined retention of the rights and control over the farm assets is more deeply rooted in the desire to secure 'family continuity' on the land than to challenge patriarchal gender relations: 'If people didn't make an effort to carry on, there'd be no carrying on.'

Thus, for example, she is at pains to stress her 'supportive' role as wife and mother and to play down both her technical competence and managerial control, mindful of local and family 'opinion'.

> I don't call myself Dr so and so . . . I try and, you know, just really be part of a family [the conjugal family household]. Most people here don't know I have any education at all. Sometimes the family [her siblings and their husbands] are a bit bitchy about it, but then they always lean on you when they're in trouble.

Moreover, her ambition is still to pass on the family farm intact to her eldest son.

The Browns at Castleton: a transitional farm (type B/C)

The Browns moved to their farm on a County Council smallholding estate in Surrey in 1967, following a number of years managing pig farms in Bedfordshire.[7] They originally had the tenancy to eight acres of land, but this was increased to 17 acres in 1976, when a neighbour retired and the Council allowed them to amalgamate the holdings. In addition, they used to operate another farm in Colne where they employed a farm manager, but following a major financial collapse in 1979 they relinquished the lease. Since 1983 they have rented 25 acres of pasture land, on an annual basis, for grazing cattle. Currently the business involves two households, Mr and Mrs Brown, who are in their late 40s, and who live on the main holding, and their second son, Alan, who lives in a separate household with his girlfriend near Colne. This is illustrated in Figure 7.10.

Mr Brown is the sole tenant on both pieces of land, but the business assets are now owned by Mr and Mrs Brown and their son in partnership. A limited company was first set up to control the farm assets in 1979, following the financial collapse of the farm, taking the form of a partnership between Mr and Mrs Brown, as a means of protecting their personal income and assets in the event of liquidation. In 1984 they set up a second company to cover their new enterprises and to give their son a stake in the holding in recognition of his major labour contribution to the farm during very bad times. Both companies now involve all three family members as partners, with Mr Brown holding a majority stake and acting as business principal.

Key

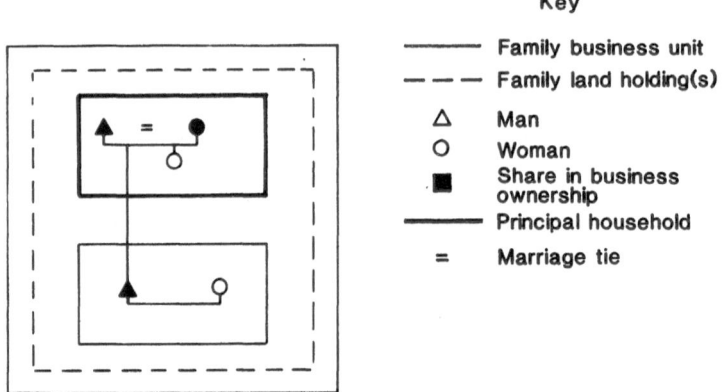

	Family business unit
— — —	Family land holding(s)
△	Man
○	Woman
■	Share in business ownership
——	Principal household
=	Marriage tie

Figure 7.10   Household structure at Castleton

Between 1979 and 1986 the farm was operating with a large debt burden which nearly forced the liquidation of the business in 1983. This debt has been the main influence on the development and structure of the current farm labour process, and specifically, the low levels of commoditisation of the non-agricultural production and household consumption processes. The origin of this debt burden lies in a period of overborrowing to support the expansion of pig production to a second holding in 1978. This expansion coincided with the imposition of new Ministry regulations controlling pig feeding systems and of new County Council restrictions on the sale of produce not grown on the holding itself.[8] The business was unable to service the loan and the family retracted production to the core holding, and have been trying to pay it off ever since.

The farm labour process comprises three circuits, domestic household labour, agricultural labour and non-agricultural farm labour, the latter representing the major source of income and profit on the farm. The major agricultural enterprises include 2000 battery hens for egg production; 60 beef cattle and 70 veal calves; 150 pigs; 600 turkeys (at Christmas time only); and pick-your-own raspberries and strawberries on about five acres. The non-agricultural circuit is centred on a meat processing enterprise producing 'home-made' pies and meat products, and a retail butcher's shop on the holding set up in 1981. This enterprise produces about 500 pies and quiches per week in peak season (summer), together with hams and Scotch eggs. Almost all the farm produce is now sold through the shop in a prepared form, whether as butchered meat or processed food, and only surplus is sold to the wholesale market. Until 1984 the domestic labour circuit incorporated the care of three teenage children. It currently extends to servicing the hired labour, in the form of the washing and ironing of working clothes, and to the care of elderly in-laws. The current structure of the farm labour process and its main gender divisions is shown in Figure 7.11.

There are marked divisions both between and within the three labour circuits. First, Mrs Brown is involved principally in the shop, butchering and food processing side of the business, which also employs two full-time workers and two part-timers in the shop and Mrs Brown's mother (who is paid in kind rather than a wage) in the food production process. Both of these sectors were set up on Mrs Brown's initiative as a way of maximising value added on the farm, which is more profitable than agricultural production itself. They exploit her skills as a butcher, learnt from her father and training at Smithfield College, and extend her domestic labour skills in the preparation and cooking of meat pies, Scotch eggs, quiches and the like into a valorised sector of production. Mrs Brown is also responsible for the book-keeping for the whole farm business and for turkey production in the Christmas season. In addition she is solely responsible for all the domestic labour in the household, except for occasional help from her daughter, who still lives at home.

Mr Brown and his son work solely in the agricultural labour circuit, Mr Brown specialising in the fruit crops, pigs and battery hens, and

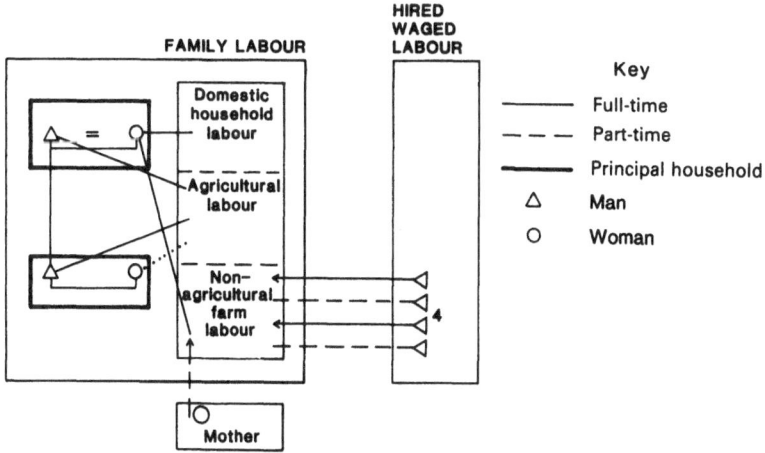

Figure 7.11   The labour process at Castleton

his son running the beef enterprise. Since their financial collapse in 1979, no hired labour has been employed in the agricultural circuit, but the highly seasonal nature of many of the production processes means that it still relies heavily on casual help from a wide circle of family members whose labour is incorporated for specific tasks, such as turkey plucking at Christmas and supervising the pick-your-own (PYO) enterprise during the summer season. This labour, like that of Mrs Brown's mother in the food processing sector, is 'paid' in kind, that is through the extended circulation of subsistence goods from the farm, but relies largely on a sense of family loyalty. It is only in the shop that regular full-time hired labour is employed, to do the butchering.

The conditions of labour also vary between the circuits, and particularly along gender divisions. Mrs Brown's food processing operation, for example, is poorly equipped, using a £25 second-hand domestic oven from a local charity shop to produce some 500 or so pies and so on a week. It therefore relies heavily on the intensive labour of herself and her mother. The agricultural production processes, by contrast, are technologically intensive (excluding the cattle) and it is in these sectors that Mr Brown's labour is concentrated. The labour of the three core family members is commoditised alongside that of the hired workers, as technically they are all employees of the two companies. However there is an unequal distribution of wages between the family members, with their son, Alan, getting most, then Mr Brown and then Mrs Brown. Alan earns the most in recognition of his mortgage commitments and to sustain his interest in the holding in a situation where his brother and sister both earn considerable salaries working in the City. Mr and Mrs Brown have a household income from the holding of £150 per week of which she gets £37 (the maximum allowable before her husband's income is taxed) and he gets the balance. Figure 7.12 shows the farm budget for the labour circuits on the farm.

The turnover of the farm is about £200 000 a year, of which nearly £188 000 is derived from sales through the shop. Thus, whilst it is Mrs Brown's areas of responsibility and labour in the business which generate the greatest amount of value-added and the largest share of money-income, she is the lowest paid family worker. Moreover her wage is the staple 'housekeeping' budget in the Brown household and only 'extras', such as going to the cinema or new items of clothing, are purchased from her husband's wage for collective household consumption. A major feature of the budget structure of the farm,

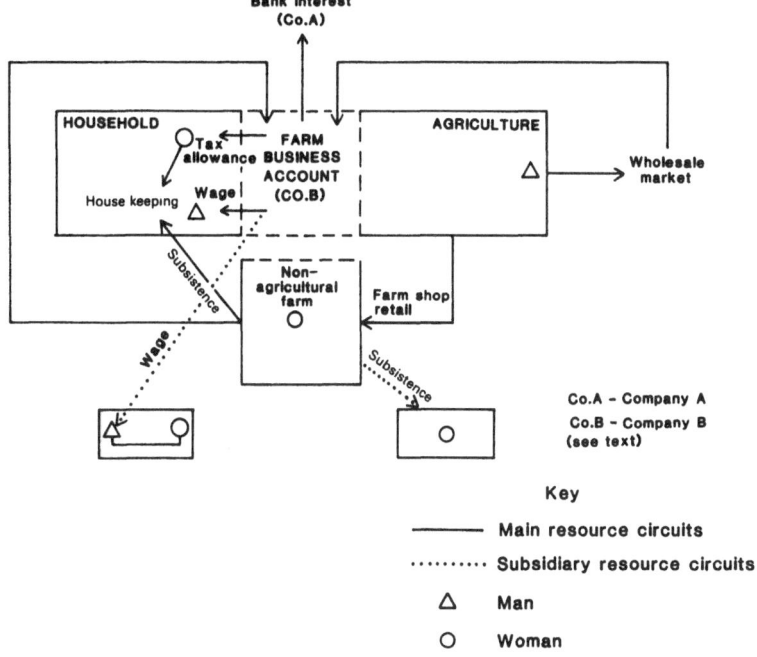

Figure 7.12   Budget structure at Castleton

especially in view of the drain on money income from the business debts, is the circulation of subsistence goods from the farm to the household consumption process. Thus the household is self-sufficient in meat and has access to fruit and vegetables through an informal exchange arrangement with various neighbouring smallholders who are 'growers'. In addition, as with all the cases of family farms organised as formal business entities, the household consumption process is subsidised via the business account through the payment of utility bills (and a range of more and less legitimate items) which attract tax relief through the business account.

FAMILY FARMING UNDER STRAIN

The Watsons and the Evans represent family farms at opposite ends of the spectrum in terms of the level of commoditisation of the agricultural production process. However the analysis suggests remarkably similar processes at work in the breakdown of the unity of household and enterprise central to domestic commodity production. In each case, debts to banking capital incurred through household and enterprise reproduction processes are shown to have undermined the integrity of the household production system and to have reshaped the gender division of labour around an increased dependence on off-farm waged labour. While in both cases the income earned from such work is central to the household's livelihood and, moreover, subsidises the agricultural production process, its use and meaning is shown to be markedly different, depending on whether it is husband or wife who provides the waged-income. The analysis highlights the contradictions raised by these external dependencies for patriarchal gender relations and the domestic political economy of these farm households.

The Watsons at Vale Farm: a marginal farming enterprise (type A/B)

The Watsons have been farming in the Menston area in Dorset since 1919. Mr Watson's father first became tenant of the 50 acres at Vale Farm in 1965 but in 1972 the owner died and he bought the freehold to the holding. Mr Watson took over the occupancy of the house and land in 1978, when his father died, but has been 'renting' the land since then from his mother who was left it in his father's will and has refused to sell it to him. His brother is also farming part-time at Eden on land purchased by his father in anticipation of the Menston by-pass being built through it. Mr Watson, his two brothers and mother are now partners in the land at Eden, but are still waiting for the by-pass. The household occupying Vale Farm currently consists of Mr Watson and his wife Jill, who were married in 1980 and are both in their early 30s and their two small children. The household structure is shown in Figure 7.13.

During their period of occupation, the agricultural production system based on family labour has broken down, as it has been unable to support the household or to reproduce itself. This decline was hastened by the increasing debt burden of the farm enterprise and the growing dependence of both household consumption and agricultural reproduction on external loan capital from one of the major clearing banks, which eventually threatened to foreclose on the loan. Mrs Watson has played a dual role in this process. On the one hand, she is not from a farming background and, despite assisting

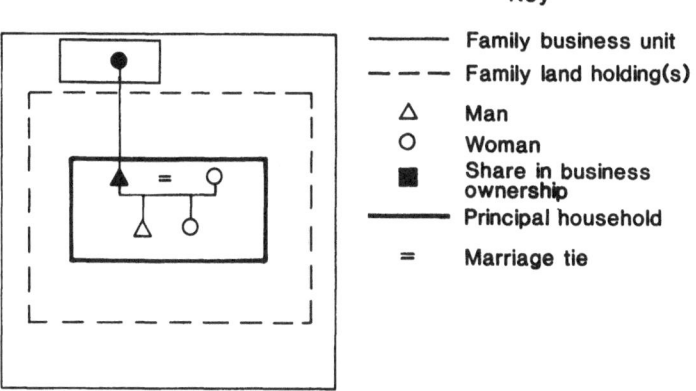

Figure 7.13   Household structure at Vale Farm

with some of the labour tasks associated with the sheep, has no experience or training in agricultural work. As a result, she has been unable to contribute extensively to the agricultural labour process which previously had been dependent upon family labour supply. On the other hand, it has been Mrs Watson's waged labour off the farm as a supply teacher which, from the early days of her marriage, has subsidised the agricultural production process, sustained the household income and kept the bank at bay.

Since their marriage both Mr and Mrs Watson have set up their own small businesses on the farm in order to supplement household income. As a result family labour has been withdrawn from the agricultural circuit of the farm. Farming itself has been run down from a fairly intensive beef and vegetable production system to a labour-minimising strategy consisting of a flock of 50 breeding ewes producing about 50 lambs for meat each year and which 'more or less look after themselves' and 'keep the land tidy'. Until 1986 the bulk of the land was used to produce hay, with 10 acres cultivated for barley for animal feed. However, as their work in non-agricultural enterprises on the farm has grown, hay too has proved too labour-demanding and from 1986 onwards all the land was cultivated; 20 acres in spring barley, 20 acres acres in winter wheat and 10 acres in rye, using contract labour. The current farm labour process thus consists of a domestic labour circuit, a non-agricultural on-farm labour circuit, an off-farm waged labour circuit and only a marginal agricultural circuit. The structure of the farm labour process is

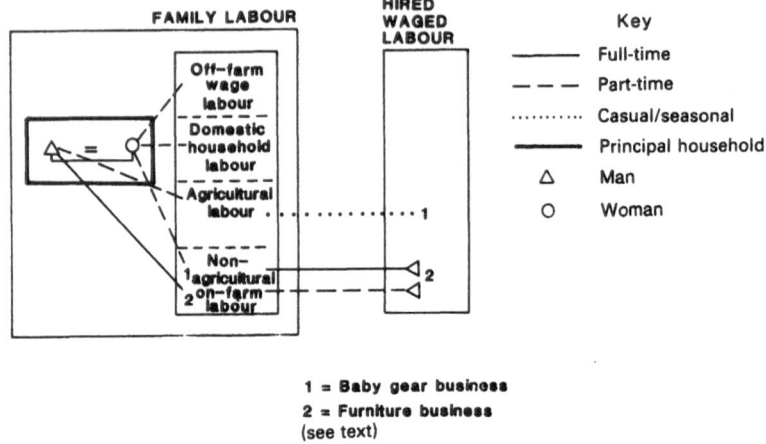

1 = Baby gear business
2 = Furniture business
(see text)

Figure 7.14   The labour process at Vale Farm

described in Figure 7.14, indicating the major gender divisions within and between these circuits.

Mr Watson still provides most of the labour in the sheep enterprise on a highly seasonal basis, with Mrs Watson's assistance at lambing. The change to sheep was prompted by the European Economic Community (EEC) ewe subsidy on which the profitability of the enterprise is entirely dependent. The sheep produce both wool and lambs for meat. The change from hay to cereal cultivation has been undertaken on the basis of making use of contract labour. This not only releases Mr Watson from these labour tasks which he no longer has the time to do, but also makes use of the contractor's specialist machinery which the Watsons had to sell to raise money income. In consequence, the profit from these enterprises is almost entirely absorbed by the cost of hiring the contract labour (at £3.50 per hour, 1986 prices) and serves to reproduce the agricultural production process itself rather than generating a source of income for the household or for investment in the non-agricultural enterprises on the farm. There are two such enterprises on the farm which make use of redundant farm buildings. These are a furniture restoration business run by Mr Watson and his neighbour and a second-hand 'baby gear' business run by Mrs Watson, which until recently involved a partnership with a neighbour.

The furniture restoration business was set up in 1983 as a partnership between Mr Watson and his neighbour who now work full-time on it and employs one other full-time worker. Each of the partners invested £2000 in the project and have since paid themselves £50 a week each from the business. The work involves stripping and renovating old furniture for the 'antiques' market. The business has incurred growing overheads, particularly in securing market outlets, their only regular outlet being a shop in a nearby town on a tourist route where they rent space at £50 per week and 2 per cent commission to the owner of the shop on all sales.

Mrs Watson's second-hand baby gear business was set up in 1983 when she stopped her job as a full-time supply teacher on the birth of her first child. Initially it was run as a partnership with her neighbour, using a government Enterprise Allowance. Instead of a regular wage they paid themselves a lump sum from the profits when the business could sustain it. In 1986, however, they ended the partnership and Mrs Watson now runs the business on her own. Since then it has been taking about £200 a week and, apart from renewing stock and advertising in local papers, it has no major overheads. Mrs Watson does all the book-keeping for the farm (except for the furniture business) as her husband is 'not very interested' in that side of it.

Neither of these non-agricultural on-farm enterprises provides the household with a reliable living and, more significantly, neither has been able to pay off the large debts previously generated by the agricultural business. By 1983, the farm business account had an overdraft of £22 000, which required £150 per month just to pay off the interest. Mrs Watson's full-time supply teaching between 1980 and 1983 serviced this growing debt, but she gave this job up after her first child was born. None the less she has reluctantly had to return to supply teaching part-time after having her first child, and again, more recently in 1986 after the birth of her second child in order to secure the household income and keep up the repayments to the bank. These debts became worse in 1985 as the furniture business also went 'into the red' following a slack period at a time when the joint 'household' account was also overdrawn.

The dependence of the household and farm businesses upon banking capital has profoundly influenced the labour strategy and budget structure of Vale Farm. Mrs Watson's return to supply teaching has generated the only regular source of money income with which to meet household consumption needs. It has also enabled the mounting and diverse farm debts to be restructured by the bank in the form of a mortgage on the house. This in turn attracted mortgage tax relief which immediately cut the costs of repaying the debts by a third. The current domestic budget structure on the farm is shown in Figure 7.15.

Mrs Watson's salary pays the mortgage, which in fact represents accumulated business debts from the agricultural enterprises and the furniture restoration enterprise, together with debts arising from household consumption in a period without a positive income. The agricultural production process, over and above the costs of its own reproduction through the contractor and basic inputs, pays the main utility bills of the household and non-agricultural businesses. Mr Watson still retains his £50 weekly wage from his furniture work, some of which goes towards the housekeeping to supplement the input from Mrs Watson's salary, despite the outstanding debts from the furniture business account which are now effectively being paid off by Mrs Watson's wage labour, through the mortgage. All of Mrs Watson's salary goes into the 'joint account' and the bulk of it is absorbed by the mortgage, but it is all for 'household use'.

At Christmas 1986, for the first time since their marriage, the joint

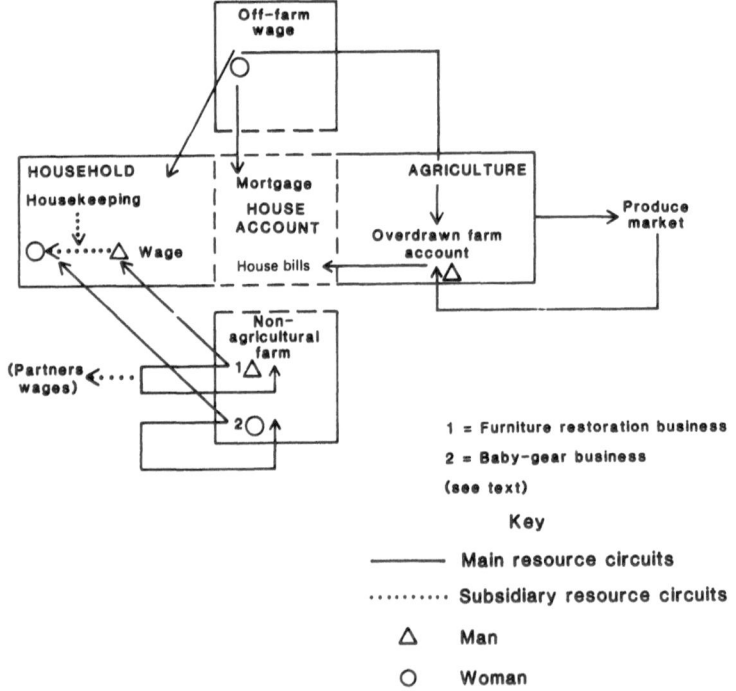

Figure 7.15   Budget structure at Vale Farm

account was in 'the black'. However, while Mrs Watson's off-farm wage labour has rescued both household and business from dire financial straits, this has not been without some personal cost to both Mr and Mrs Watson. The situation has undermined the material basis of patriarchal power in the conjugal household by reversing the gender roles of 'breadwinner' and 'dependent' and thereby challenging the ideologies, practices and gender identities associated with them. Mr Watson, for example, tries to discourage his wife from developing her business further, arguing that, '. . . if you keep it as it is it's ideal . . . your problems start when you become more ambitious and try to make a living out of it'. His attitude conveys tensions bound up with his identity as a 'man' in the local farming community, where he is judged by his ability to earn a living at farming.[9]

Equally, Mrs Watson is unhappy at having to do supply teaching when she feels she should be at home looking after the children. Despite her off-farm wage-labour and work in the baby-gear business, she still does the bulk of the domestic labour, particularly housework and cooking, both because she 'feels she ought to' in order to be a 'good wife and mother' and because Mr Watson is reluctant to take on such tasks, as this would compound the tensions of 'role reversal' already expressed in financial terms. As with the Prices at Fountain Farm, although realised in a different form, the increased economic power of Mrs Watson does not simply erase or contradict the ideological and material basis of the patriarchal gender regime which defines her position as a 'farm wife', but rather transforms its everyday practice and meaning for the individual men and women involved.

The Evans at Rough Farm: a share-farming partnership (type C/D)

The Evans have been living at Rough Farm near Wandle in Surrey since 1983, but have lived on the estate of which it forms a part since 1978, when Mr Evans was employed as manager of the estate. The estate was bought in 1978 by a property development company from a family in the industrial gentry and is held for its development potential rather than primarily for its agricultural use. However the company have sought to maximise their income from the land while it remains in agricultural use by reconstituting the estate in the form of three separate 'share-farming' operations with individual farmers. Mr Evans was made redundant as the estate manager and taken on instead in such a 'partnership' arrangement at Rough Farm. This is the smallest unit on the estate at 250 acres. The estate as a whole comprises 1120 acres, of which 190 were taken from Rough Farm in 1979 and converted into a golf course, and two other farms of 420 and 260 acres. As Mrs Evans explained, 'the arrangement with the landowner represented the only way that we could have gone into farming on our own account'.

But, while both Mr and Mrs Evans feel more 'independent' under this new arrangement, its impact upon their income has been severe. The farm operates on a family labour system which cannot sustain the level of profitability needed to meet the consumption requirements of the Evans household, the reproduction of the agricultural means of production (for which Mr Evans is largely responsible under the terms of the partnership) *and* the share of the profits appropriated by the property company under the terms of the agreement.

The structure of the Evans household in relation to the property rights over the land and business assets of the farm is set out in Figure 7.16. It comprises Mr Evans and his wife Vicky, who were married in 1976, and their three young children. Both Mr and Mrs Evans are in their early 30s and neither came from a farming family, although Mr Evans's father managed an intensive livestock unit in Shropshire.

The business is organised as a 'share-farming' arrangement between the land-owners and Mr Evans, in which he is responsible for providing the working capital and labour for the agricultural production process. The profits are split between the property company and Mr Evans on a 30–70 ratio. The property company is responsible for the maintenance of the farm house and the land. This arrangement

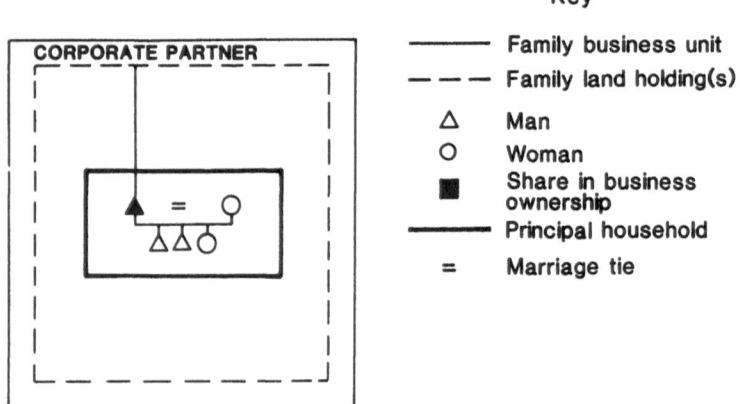

Figure 7.16   Household structure at Rough Farm

avoids the loss of vacant possession (and hence protects the market value of the land) whilst providing a commercial income for the company from its current agricultural use. Mr Evans invested his £4000 redundancy pay into the business, but has since become increasingly heavily indebted to a variety of banks and merchants for the purchase of basic agricultural inputs and for household consumption, as the farm labour process has not been producing enough money-income to sustain either itself or the household's needs. In this context the Evans have been rapidly diversifying their farm labour activities in order to increase household income which is now entirely derived from non-agricultural sources. Any profits from the agricultural labour process are ploughed back into the farm business, either directly through investment in new inputs or through the payment of the profit share of the corporate partners, or indirectly via the payment of interest to loan capital.

The farm labour process currently consists of an agricultural and domestic labour circuit, a non-agricultural farm labour circuit and a seasonal (winter) off-farm wage labour circuit. This is depicted in Figure 7.17. The agricultural circuit is almost entirely arable, with some oats and wheat grown on contract for Jordans and Quakers and barley for commercial animal feed. A major commercial crop is also hay and barley straw which fetch very high prices locally from the 'horsey people' who live in the area. In 1987 Mr Evans also agreed to try 10 acres of sunflowers on contract for a margarine production company and will expand this crop if it is successful. In addition, there is an extensive range, but limited number, of livestock kept on

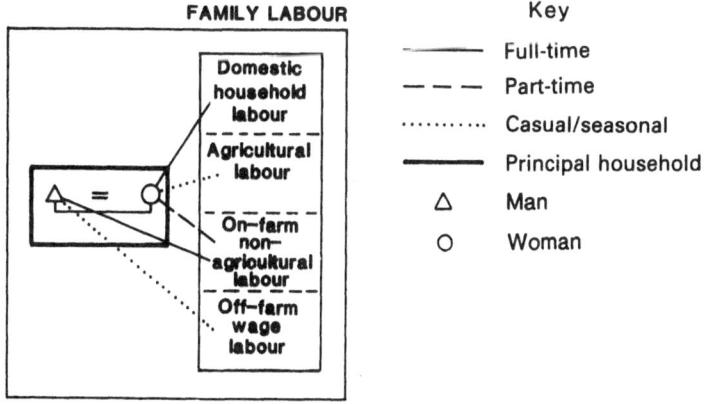

Figure 7.17   The labour process at Rough Farm

the farm for small-scale commercial and subsistence purposes. These include 30 ewes and lambs, 20 hens for eggs, 25 ducks and 20 geese at Christmas, and 15 beef cattle. Within this labour circuit there is a marked division of labour, with Mr Evans doing all the arable work and Mrs Evans doing most of the livestock work. Mrs Evans is responsible for all the domestic household labour, which includes looking after two small children and, recently, a new baby, the preparation of meals and all the housework and laundry activities.

The non-agricultural farm labour circuit has expanded as the Evans's income has fallen and their debts have risen. This now includes five 'do-it-yourself' liveries (letting out stables and providing straw for local horse-owners); renting out a barn to a local builder for storage; educational farm tours by local school parties and, most recently and significantly in terms of income, a farm shop selling produce brought in from markets and growers whom they know. All these activities are farmyard or farmhouse based and involve mainly the labour of Mrs Evans, except for the farm tours and the purchasing of the shop produce. In addition, Mr Evans took a full-time waged job as a garage assistant at the local garage in the winter (1986/7) to generate income for household consumption over a very bleak period, a pattern likely to be repeated. Mrs Evans is a trained radiographer who gave up work on having her first child. She is contemplating going back to work, although neither she nor her husband thinks that it is 'right' for mothers not to look after young children full-time. None the less their financial situation might well force them to reconsider the gender division of labour in the future and oblige Mrs Evans to earn a regular income by practising her professional skills again as a radiographer.

As Figure 7.18 shows, the household budget is now entirely split. On the one hand, the product of the agricultural labour process circulates almost entirely through external market links, servicing business debts, the partners' share of the profits and reinvestment in new agricultural inputs. On the other, household consumption which is also highly dependent on money-income is funded through the non-agricultural farm labour circuit and Mr Evans's off-farm wage labour in the winter. In this sense, the productive and reproductive relations of the household production system on which 'family farming' is based have broken down and the position of the Evans household is much closer to that of a salaried worker, despite the fact that Mr Evans is technically no longer a farm manager.

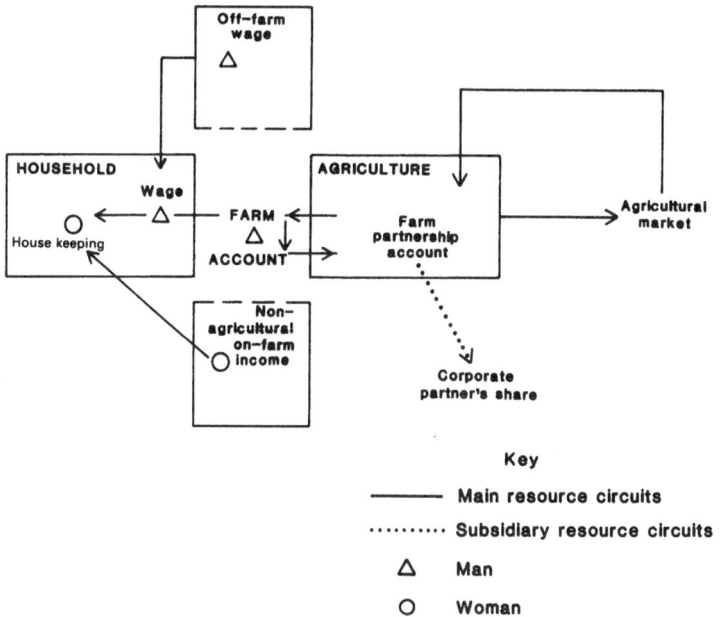

Figure 7.18   Budget structure at Rough Farm

The income from the farm shop now pays the housekeeping money, which is managed by Mrs Evans for domestic household consumption. All the farm activities, bar the liveries, the geese, ducks and eggs, go straight into the farm business account, from which the partners and banks are paid and new inputs purchased. The farm account is not yet self-sustaining, however, because the highly seasonal nature of income from arable products means that the purchase of new inputs is still dependent upon borrowed money. The situation has been made worse by the bank cutting the business overdraft allowance by a half to a maximum of £5000, which meant that debts of £6000 to the inputs merchant had to be paid off immediately. This demand cleaned out the farm business account following the 1986 harvest, at the time when the account should have been most in credit. It was as a result of this change in the bank's lending policy that Mr Evans took on his winter off-farm job to finance household consumption and the purchase of new inputs for the next agricultural cycle on credit from the bank. Mrs Evans's labour in the livestock enterprises mostly feeds into the farm account as well, except for the 'egg money' which is kept to buy chicken feed and as an emergency household fund. Unlike her husband, however, Mrs Evans is not paid a monthly salary from the business account. Mr Evans's salary now goes to paying for agricultural inputs, rather than for personal or domestic consumption.

8 Conclusions

The central concern of the book has been the reconstruction of the political economy of family-based production from a feminist perspective. The reconstruction I have proposed has a number of dimensions and implications for the meaning and practice of political economy. The most obvious is its incorporation of a theory of patriarchal gender relations. Just as important, it builds on two other themes already emerging in the literature on agrarian change, and well established on a wider canvass. The first is the development of a post-structuralist or humanistic, interpretation of Marxist political economy; an interpretation concerned with social relations as lived relations, actively constituted through the processes of subjective meaning and interaction (Sayer, 1984). The second is a broadening of horizons from a narrow focus on the process of production, as capital accumulation, to the processes of consumption and livelihood.

This final chapter is organised around these three themes. Its purpose is to move beyond the particular people and places under scrutiny and to draw out the broader significance of the theoretical arguments which underpin my interpretation. I look first at the main implications for current debates and future research about the process of contemporary agricultural restructuring and the status of the family farm. The second part of the chapter turns to a consideration of the relevance of these themes for wider debates raised in the introduction about the relationship between gender and class and the development of political economy analysis.

RETHINKING FAMILY FARMING

My initial contention was that the invisibility of women in conventional theories of family-based production is bound up with their failure to grasp the distinctiveness of this group of producers. From the subsequent analysis of the case of family farming three linked weaknesses can be identified in these conventional approaches, which articulate this problem more precisely:

1. A narrowly productionist concept of labour restricted to the process of agricultural commodity production;

2. A chaotic concept of 'the' family lacking a critical theory of the
    social relations which structure it;
3. A failure to theorise gender relations.

In seeking to resolve this problem, the adequacy of orthodox
concepts of simple, independent and petty commodity production has
been called into question as the basis for analysing family-based
production. Such concepts retain a restrictive focus on the social
categories of capital and labour and fail to specify the distinctive
features of the social relations of the family that characterise their
object of analysis. These concepts share the central premise that such
forms of production are unitary and lack a 'class relation'. Such a
premise is simply not sustainable in the light of the volume of
evidence presented here and elsewhere of the systematic gender
inequalities which structure family farming. As we saw in Chapter 3,
domestic commodity production has been advocated as a better way
of conceptualising the distinctiveness of family-based production.
This term highlights the unity of family household and commercial
enterprise and gives greater theoretical weight to the significance of
kinship and household relations. I have sought to develop these
arguments beyond the terms of current debates by tracing the ways in
which gender relations intimately structure the organisation of family
labour and capital.

A case has been made for incorporating a theory of patriarchal
gender relations into political economy analyses in such a way
that they can be extended to make sense of the internal relations
of family-based production. Such an approach provides the
corner-stone for a more comprehensive *domestic* political economy
framework. The conventional 'constraints' versus 'resilience' debate,
reviewed in Chapter 2, has led to an increased recognition of the
significance of gender to family-based production and to various
attempts *empirically* to 'add women in'. The feminist reconstruction
put forward here suggests one way through the *conceptual* impasse
created by the terms of that debate for the analysis of gender. Gender
relations and the subordination of women as 'wives' become central
to an understanding of the labour relations peculiar to family farming.
Family farming gains coherence as an analytical category centred on
patriarchal family labour and property relations but covering a range of
actual regimes in which kinship, household and enterprise intersect in
different ways at different levels of commoditisation in specific local
contexts. These two points are developed in turn below.

The labour process

The analysis of the farm labour process, particularly the detailed case studies in the last chapter, supports arguments elsewhere that the categories of productive and 'non-productive', or reproductive, labour are conceptually problematic and analytically unhelpful. Four labour circuits have been distinguished. These comprehend all the labour activities sustaining the interwoven processes of production and reproduction by which household and enterprise are bound together and a livelihood is made from the land. The relative importance of these circuits is historically variable, as are the relations under which they are carried out. This framework permits the analysis of hired as well as family labour. It highlights the significance of gender and, potentially, other social divisions in the farm labour process within and between these circuits.

However it is also clear from the case studies that a non-commoditised off-farm circuit of labour, drawing on wider kin and community networks of reciprocal labour exchange, is of greater significance than first thought. They are likely to vary sectorally and regionally but, in general, the extent of such networks seems to decline with increasing commoditisation. Equally important, the notion implicit in much work on the family labour process reviewed in Chapter 2, that familial relations are by definition non-commoditised, has been challenged here. Both the survey and case study material demonstrate an increasing incidence of individual remuneration for family members' work with increased levels of commoditisation.

The family labour process has been shown to be organised around a gender division of labour structured by the patriarchal institution of the conjugal houschold and a gender division of property rights structured by patrilineal kinship practices. Through these relations control over capital, the technological means of production, and income from valorised labour processes is concentrated in the hands of men. The family labour process can thus be defined more accurately as a patriarchal labour process.[1] This is not to suggest that all family-based production everywhere is structured by the same patriarchal relations. Nor does it imply that gender relations are *necessarily* patriarchal in all such forms of production. But such a definition is useful in so far as it draws attention to the centrality of gender in structuring the labour process and to the structural position of women as wives across many forms of family-based production.

The labour process emerges as a key site for constructing and contesting patriarchal gender relations across a range of structural conditions and specific circumstances. Women's subordinate position in the domestic political economy of the family farm can be seen to be legitimised through familial gender ideologies which naturalise gender inequalities. Chapter 6 examined in detail some of the ways in which the gender role of 'wife' builds upon, and reinforces, more widely constructed subordinate gender identities of femininity and womanhood. These roles and identities are not fixed but are transformed both in the commoditisation process and by struggles for women's emancipation. The ideological processes which both inform and find expression in the gendered structure of the labour process are clearly very complex. They implicate women as well as men in the legitimisation of patriarchal production regimes and in the defence of traditional practices. However patriarchal gender ideologies and practices are not a simple function of the commoditisation process. A number of tensions between them are apparent, such as in the development of more corporate forms of 'family' property. Such tensions are experienced by women as a conflict between their individual and familial identities and create space for the renegotiation of status around key issues such as labour remuneration and property rights.

But, as Chapters 6 and 7 also make clear, patriarchal gender relations are not sited exclusively in the labour process, nor are the ideologies mobilised in family labour practices solely constructed there. This raises two important limitations for the kind of analysis that I have undertaken. One is that, in focusing on the regime of the family household, the attention paid to the wider gender order as it influences that regime has been restricted. In particular, the significance of peculiarly 'farming' gender ideologies, associated with an occupationally defined community (Strathern, 1984), and of locally specific gender ideologies and practices (Bowlby *et al.*, 1986) has not been explored.[2] A second is that the material relations of sexuality, and emotional and physical violence, which are constitutive of patriarchal relations, require more direct and sustained attention than they have received here. In conjunction with the untreated issues of generational relations within the family household and 'race', as a social division cross-cutting gender and class, the analysis presented is clearly incomplete. However these limitations point to areas for future research for which the domestic political economy framework put forward might serve as a useful starting-point.

Commoditisation and differentiation

Farming Women explores the substantial differentiation amongst family farming regimes in terms of their internal social relations and endorses an interpretation of commoditisation as a variable process. In this it takes up the arguments, examined in Chapter 2, against the categorical logic characteristic of orthodox agrarian political economy concerned with specifying hard and fast boundaries and laws of motion for discrete theoretical categories like SCP. Such modes of theorising construct family farming as if it were a homogenous class, whether conceived of as 'exploited outworkers' or as 'self-exploiters'. Along with other recent critics, I have sought to move towards a mode of theorising concerned with the processes by which the rich variety of family-based production is constituted and transformed in particular contexts. Rather than equating family farming with the concepts of PCP or DCP, it can be seen to represent a spectrum of forms of family-based production within which DCP is identifiable as one form, centred on family labour.

This approach is developed here in two main ways. Firstly, commoditisation has been shown to affect family farms in terms not just of the organisation of agricultural production but of the relations of household and enterprise reproduction, and the familial ideologies and value systems which inform them. An analysis of this complex pattern of differentiation somehow has to bring these dimensions of the commoditisation process together. While by definition a simplification, the typology used here does assist in this task. It permits the shift from labour to capital, as the organising principle of the internal social relations of the family farm, to be related to the development of commodity relations with the wider market economy. The detailed case studies in Chapter 7 provides a vivid illustration of how a variety of specific forms of family farming, at different levels of commoditisation, work in practice.

The second aspect concerns the ways in which capital builds on pre-existing gender divisions and relations in the family farm and re-shapes them in the commoditisation process. Once again, this is an empirical process, not a logical one, which takes place unevenly and through human agency within specific local contexts. Here we have seen how women become marginalised from key positions in regimes centred on family labour to peripheral positions within regimes centred on family property. This pattern of 'defeminisation' is clearly only one of a range of possible outcomes, but a more fundamental rethinking of the significance of gender is necessary to make sense of it. Rather than being relegated to a dimension of the relations of reproduction, or to an ideological gloss on the production process, gender emerges as a deep-seated feature of the production process itself.

In this light, the preoccupation in the established literature with commoditisation as the sole, externally determined, force for change in the structure and viability of the family farm appears very partial. Structural tensions in the patriarchal gender relations of family farming can be seen to form the basis of another, potentially transformative process. Despite women's highly individualised experience of work as farm wives, and the complex ideological processes legitimising patriarchal labour relations, signs are emerging of more collective action by farm women to improve their position.[3] Continental Europe and North America have both seen farm women, particularly younger women, organise to fight for more equal property and inheritance rights over the farms where they spend their working lives. While less radical than some of their continental counterparts, the Women's Farm and Garden Association and the Women's Farming Union in Britain are both active in such campaigns.[4] Moreover such campaigns have been instrumental in getting the Parliament of the European Economic Community to adopt a directive promoting the equal treatment of men and women, including the constitution of individual rights for women working in family businesses (86/613/EEC). As a consequence even the National Farmers' Union, the powerful but highly conservative farming lobby in Britain, has been forced to address these issues.[5] These are small beginnings, but are indicative of a transformative potential as significant for the future of family farming as the process of commoditisation.

SHIFTING PERSPECTIVES

Agriculture has in the past been marginalised from mainstream social and economic research in advanced capitalist societies because its family-based relations of production have been held to be 'residual' and atypical in comparison with other branches of modern industry. This misleading demarcation has been challenged here. Several authors, such as R. Williams (1984) and Benvenuti (1985), have already noted that, with the growth of various forms of non-wage work and non-corporate production elsewhere in advanced societies, family farming has gained a wider sociological significance. At the simplest level one can extrapolate key arguments from this study, such as the gendered structure of family property and labour, to family-based production in other sectors of the economy. But even apparently straightforward extrapolations of this kind require caution. While in principle the family farm may be assumed to be like any other family enterprise such as a shop or workshop, the biological nature of the farm production process makes it distinctive in so far as it implies the *potential* for direct subsistence from the labour process. To ascertain the real significance of this feature would require a comparative analysis of family enterprises in different sectors of the economy.

The kind of feminist reconstruction of family-based production proposed here offers new perspectives on various enduring sociological issues such as the 'uneasy stratum' of the small business and the petite bourgeoisie (Bechhofer and Elliott, 1981). It also casts doubt on the innocence of everyday terms such as 'the self-employed' which, like 'the farmer', construct their subject as autonomous individuals rather than as rooted in a complex network of unequal family and household relations.

However my concern in this final section is less with extrapolating this framework to similar situations outside farming than with its broader contribution to current debates about the place of gender in theories of social stratification and economic restructuring (Crompton and Mann, 1986). These debates have thus far been informed primarily by analyses of the wage-economy (Hart, 1989). From the perspective of the family farm, they appear clouded by a divisive spatial imagination which locates family and economy, patriarchy and capitalism, feminism and Marxism in separate and opposing domains. In contradicting this imagination, such agrarian social structures demand, and inform, a wider shift in the conceptual landscape.

Gender, class and livelihood

In arguing for a theory of patriarchal gender relations based on a conceptually autonomous sex-gender order, *Farming Women* picks up two persistent criticisms in current debates. As Cockburn notes, this interpretation of patriarchy is often claimed to be ahistorical and to leave the theory of capitalist class relations untouched, thereby abandoning the economic sphere to Marxism (1986, p. 82). There is no doubt that some conceptualisations of patriarchy have been highly ahistorical (and ethnocentric) and that the process of abstraction requires more rigour and care. But, as Cockburn goes on to point out, only by disarticulating sex-gender order and mode of production can we make sense of patriarchy as an historical process with very different dynamics to, and reciprocal effects upon, capitalism in particular times and places. On the second point, she argues that the problem lies in the unidimensional way in which the sex-gender order has been conceived of primarily in ideological terms (1986, p. 83). I believe that the criticism of 'abandoning the economic sphere to Marxism' marks feminist analysis more deeply than her response admits. It is a problem rooted in the kind of dualistic thinking examined in Chapter 3, by which production and reproduction have been treated as separately sited and analytically discrete processes. I have tried to show that such an approach is not a necessary implication of accepting the conceptual autonomy of the sex-gender order.

Farming Women provides one detailed example of the way in which class relations and forms of production are partly constituted out of the opportunities for allocative and authoritative power created by patriarchal gender relations. The particular features of family farming throw light on two important dimensions of any shift towards a more integrative framework of analysis. The first is the need to rethink the nature and significance of 'the family' in the wider social and economic order, and the second concerns a reassessment of the contributions of feminism and Marxism to political economy.

The idea of the family as a particular site or arena of domestic activity has too long overshadowed its importance as a web of social relations centred on kinship and household practices and ideologies (Stacey, 1986). These relations mediate access to property, resources and skills and, in different ways in different times and places, they structure production, as well as reproduction, and class, as well as gender. Of course the analytical separation of family and economy is not just a figment of a divisive spatial imagination but an empirical feature of a specific historical geography associated with the industrial revolution. But this particular configuration has come to be represented across a range of theoretical perspectives as a universal reality with key concepts being shaped by, as well as shaping, that representation. Within Marxism, for example, Raymond Williams has argued that the idea of mode of production has systematically privileged the process and institutions of capital accumulation to the extent that it has effectively become a substitute for the broader bases of human social and economic activities (1983, p. 263). In so far as feminist analysis has been content to restrict itself to a 'domestic realm', it can be argued to have been hostage to the same distorted representation.

Williams goes on to propose a reorientation of political economy analysis around the concept of livelihood (ibid, p. 266). This idea has been central to the domestic political economy framework put forward here as a means of recasting the interweaving processes and complex geographies of production and reproduction. Most importantly it recasts these processes as practical and meaningful, rather than abstract and logical. The concept and analysis of livelihood requires a great deal more development if family, household and gender are to fully emerge from behind the box hedges of suburbia as part of the heartland of political economy. But, equally, there is already much to be learnt from the experience and analysis of developing countries which undermine standard orthodoxies about the social and spatial relations of livelihood. Comparative texts, like *Beyond Employment* (Redclift and Mingione, 1985) for example, suggest a particularly valuable model for taking forward the analysis of gender and class relations in ways which challenge the boundaries of 'home' and 'work'.

Such a reorientation of political economy raises important questions for the potential contributions of feminist and Marxist perspectives. These questions are embedded in wider debates about the nature of feminism and Marxism, and the relationship between them, which are clearly beyond the scope of this book. But *Farming Women* does identify some more specific issues for research informed by these debates. Feminist and Marxist analysis are not compatible in all their guises (Sayer, 1983) but 'post-structuralist' developments in both feminist and Marxist theories have been heralded as promising a more fruitful dialogue, centred on common concerns with understanding the processes of identity, meaning and representation and the transformative potential of human agency (Marshall, 1988; Mackenzie, 1989). I have adopted an approach very much in sympathy with these developments, but not without encountering a number of problems.

At the most general level post-structuralist modes of analysis have a dangerous tendency to retreat from the materialist concerns of political economy and from any contact with people's daily lives, becoming wrapped up instead in a seamless web of words about words. Kaplan identifies this retreat as symptomatic of a desire to make 'texts' stand in for actual social and political activity (1986, p. 12). Although I would accept that this problem is not endemic to such modes of analysis it does need to be addressed at the level of research practice, which is not easy. For example, the kinds of methodology required to tackle questions of identity and experience are fraught with ethical and practical problems. They entail, it has been argued here, a necessary process of re-presentation of research subjects' self-understanding because of the nature of social discourse as a source of information. This means that my representation of the working lives of the women portrayed does not mirror their own. Neither can it be read as somehow tapping a deeper level of 'reality' or 'truth'. It is limited to highlighting some of the discursive practices by which women come to represent their interests in particular ways; practices which have a direct bearing on the way individual women experience work as 'farm wives' and on the legitimation of their collective position of material dependence.

Despite these doubts and limitations, 'post-structuralist' theories do seem to offer our best hope for a more fruitful relationship between feminism and Marxism; a relationship which, as I have tried to show here, is crucial to the development of a meaningful political economy.

Appendix: Survey Questionnaire

POSTAL QUESTIONNAIRE OF FARMERS' WIVES IN
DORSET AND THE METROPOLITAN GREEN BELT

The questionnaire follows overleaf.

149

SECTION A   You and Your Family

1. When did you come to live on this farm?
   YEAR OF ARRIVAL:

2. When did you marry your husband?
   YEAR OF MARRIAGE:

3. Which age group do you and your husband fall into?

	YOURSELF	YOUR HUSBAND
25 or UNDER		
26–35		
36–45		
46–59		
60 or OVER		

4. Are/were your parents in farming?
   YES / NO

   If yes → (i)   Where do/did they farm? (county and parish)

   If no → (ii)   What occupation do/did they have?
   (a)   father
   (b)   mother

5. Do you have any other close relatives on *your* side of the family in farming? (e.g. brothers, sisters, cousins, uncles or aunts)
   YES / NO

   If yes → (i)   What relation(s) are they to you?

   (a)
   (b)
   (c)
   (d)

   (ii)   Where do they farm (county/parish)?

   (a)
   (b)
   (c)
   (d)

6. What kind of school did you last attend?
   (a) secondary modern/comprehensive
   (b) grammar school/high school
   (c) village school
   (d) private day school
   (e) private boarding school
   (f) other – Please specify

7. At what age did you leave school?

8. Did you do any of the following?

   PLEASE SPECIFY TYPE OF
   COURSE/QUALIFICATION/TRAINING

   (a) Higher Education.
   (b) Vocational training
       (non-agricultural)
       e.g. doctor, nurse,
           secretarial.
   (c) Vocational training
       (agricultural).

9. Do you have any children?
   YES / NO

   If yes → (i) How many children do you have?

10. What age and sex are your children?
    Boys/Men   *Girls/Women*

11. How many of your children still live at home with you?

12. If you have adult children:

	YES	NO
(a) Are any of them involved in your farm business?		
(b) Have any of them their own farm business?		
(c) Are any of them involved in farming or someone else's farm businesses?		

13. If you have younger children, do any of them want to go into farming?

14. Do you have any other relatives (other than your husband and children)
     living in your home on a regular basis?

     YES / NO

     If yes → (i)   What relation are they to you?

              (ii) What age(s) are they?

SECTION B   Your Working Day

15. How long is your active/working day (from when you get up to when
     you go to sleep) in hours?

     (a) on weekdays:

     (b) at weekends:
          (if different from (a))

16. Would you say that you spend most of your active day on activities
     (tick as appropriate)

	weekday	weekends
(a) in the house?		
(b) on the farm?		
(c) off the farm?		

17. On the domestic front, which of the following activities do you do?

	YES	NO
(a) childcare (including taking children to and from school)		
(b) preparing family meals		
(c) housework		
(d) washing/laundry		
(e) family shopping		

18. Are any of these domestic tasks shared with other family members (e.g.
     husband, older children, parents etc)?
     Please specify who it is that helps you in each task.

	Regularly	Occasionally	Never
Childcare			
Preparing meals			
Housework			
Washing/laundry			
Family shopping			

19. Do you have any paid or unpaid non-family help with any of these tasks? Please give brief details.

20. Do you have any areas of farm work which are primarily your responsibility?

    YES / NO → if yes, please give brief details.

21. Which of the following farm activities are you involved in? (tick as appropriate)

	Regularly	Occasionally	Emergency only	Never
(a) office work/book-keeping				
(b) dealing with callers/ telephone calls				
(c) running errands for the business				
(d) manual farm work (e.g. feeding animals)				
(e) helping at harvest time				
(f) dealing with employees				
(g) providing accommodation for employees				
(h) providing meals for employees				
(i) day-to-day management decisions				
(j) long-term management decisions				
(k) farm shop/gate sales				
(l) bed and breakfast accommodation				
(m) commercial horse-riding activities				

22. How many hours per week (roughly, on average) would you say that you put into the activities you have listed in questions 20 and 21 above?

    (a) Hours per week.

    (b) As a proportion of total FAMILY labour (i.e. all the hours worked by your FAMILY on the farm, including your husband).

    less than ¼   ¼ to ½   more than ½

23. Does your involvement in these activities change in type or amount on a seasonal basis? Please give brief details.

24. Do you receive payment *in any form* for any of the activities which you do on the farm, as listed under questions 20 and 21 above?

    YES / NO

    If yes

    (a)   Which activities do you receive payment for?

    (b)   What form does this payment take?
          (tick as appropriate)
          wage/salary
          share of profits/takings
          fee or fixed payment
          payment in kind
          other (please specify)

25. Would you describe your direct input into the farm business as

    (tick as appropriate)

    (a)   None

    (b)   Emergency relief only

    (c)   Casual or seasonal only

    (d)   Regular part-time

    (e)   Regular full-time

    (f) Other (specify)

26. Has the *amount* of work which you do on the farm changed since you first came to it?

    (Please tick)

    No   gone up   gone down   fluctuated

    What are the main reasons for this change?

27. Has the *type* of work which you do on the farm changed since you first came to it?

    YES / NO

    If yes → Please give brief details of changes, and why they occurred.

28. Do you *currently* have any *paid* off-farm employment?

    YES / NO

    If yes → Please give a brief description of your employment.

If no → Have you *ever had* any paid off-farm employment (including before marriage/coming to this farm)?

YES / NO

If yes → Please give a brief description *and* the reasons for stopping it.

29. If you have any income from farm activities or off-farm work, does this income (if used for more than one purpose please tick all appropriate options):

    (a) contribute to the total *household* budget;

    (b) contribute to the *farm business* account; or

    (c) do you keep it separate to spend on yourself?

SECTION C   Your Land and Business Interests

30. Do you have any legal and/or financial interest in the ownership of this *farm business*? (For example, are you a partner, shareholder or director of a family farming company, or do you have any of your own money tied up in it?)

    YES / NO

    If yes → please describe briefly.

31. Do you have any legal and/or financial interest in the ownership of any of the *land* farmed by this farm business (if different from 30 above)?

    YES / NO

    If yes → please describe briefly.

32. Have you ever *been given or bequeathed* any farm land or farm business assets?

	YES	NO
Land		
Business assets		

    If yes → (a) who gave/bequeathed this to you?

    (b) when did you receive it?

33. What have you done with any land or business assets which have come into your ownership? (tick as appropriate)

    (a) sold it

    (b) integrated it into this farm business

    (c) let or leased it out as a source of income

    (e) other (please specify)

34. Do you have any private sources of income (other than from paid employment listed in 28 above) independent of your husband and/or this farm business?

    YES / NO

    If yes, please describe briefly the source(s) of this income (e.g. pension, life assurance policy, investment income, rents etc).

35. Have you ever invested any of this income in the farm business (including purchasing land)?

    YES / NO

    If yes → please give brief details of type of investment and when it occurred.

SECTION D   Community Involvement

36. Are you actively involved in any of the following? (tick as appropriate)
    (a)   local government (e.g. parish or district council)
    (b)   church
    (c)   school governing body
    (d)   other (please specify)

37. Do you belong to any clubs, associations or local community groups (e.g. WI, NFU, playschool groups, local charities)?

    If yes →   (i)   please give brief details
              (ii)   are there any in which you are an 'officer' or organiser
    If no → (iii)   are there any groups which you would like to join?
             (iv)   what prevents you from joining or participating?

38. Is your husband involved in any clubs, associations or local groups?

    NO / YES   (please list groups)

39. Which aspects of living on a farm do you
    (a)   like most?
    (d)   dislike most?

40. Do you feel that your overall contribution to the running of the farm is important?

    YES / NO

    If yes → what do you consider to be your most important tasks/roles?

41. Do you think that the role of farmers' wives in modern farming is recognised?

    YES / NO   (a)   by the farming community
                 (b)   by the general public

    Please explain your answers.

42. FINALLY, PLEASE TICK THE BOX BELOW IF YOU WOULD BE WILLING TO BE INVOLVED IN A MORE DETAILED STAGE OF THIS RESEARCH INTO THE ROLE OF FARMERS' WIVES IN MODERN FARMING.

Notes and References

1 Introduction

1. The re-emergence of small-scale, household-based production and alternative survival strategies has also been a theme of interest in recent work on socialist countries (Szelenyi, 1988).
2. While the sociological significance of the 'family labour farm' is not in doubt, it is less easy to gauge its economic significance in terms of total agricultural output by value or volume. However this is certainly much lower than the proportion of total farm holdings represented by such farms (see, for example, US statistics which show a high concentration of production, in terms of output, on non-family farms in the order of 70 per cent of total output (Kloppenberg and Kenny, 1984).
3. The assumption that farmers are men is prevalent in western societies (Boulding, 1980). This is not the case in developing countries (see Boserup, 1970), despite the efforts of European colonialists to impose western bourgeois gender roles on 'native' women and men, particularly through religious teaching (Moore, 1988).
4. This literature is large, but amongst European contributions the following work is notable: Gasson, 1980, 1981, 1989 (United Kingdom); Lagrave, 1983; Albert *et al.*, 1987 (France); Bauwens and Loeffen, 1983 (Netherlands); Siiskonen *et al.*, 1982 (Finland); and, in the United States, Sachs, 1983; Haney and Knowles, 1988). See also the contributions to special issues of major European and North American rural sociology journals devoted to the subject of farm women; *Sociologia Ruralis*, 1988, vol. 28/4 and the *Rural Sociologist*, 1981.
5. This path-breaking tradition can be traced to the classic study of Boserup (1970) on women and agricultural development. Notable amongst more recent work are: Beneria, 1982; Wilson, 1984; Mies, 1986; Momsen and Townsend, 1987. Moore (1988) provides a wide-ranging review of the anthropological literature on the subject.
6. Jones and Rosenfeld (1981) report 96 per cent of US farm women to be wives and Barthez (1985) suggests that the figure in France is 92 per cent.
7. The French language seems, however, to provide a richer vocabulary than the English for expressing different, specific, positions held by women in farming. For example, 'femme d'agriculteur' (farmer's wife), 'agricultrices' (women farmers) and 'co-exploitantes' (joint – husband/wife – producers) (Barthez, 1983).
8. This theme is developed by Burawoy (1979) in relation to the analysis of wage labour uninformed by a consideration of gender. More recently he has extended it to describe the 'family outworker system' of early industrial capitalism as 'patriarchal apparatuses of production' (1985, p. 93). However, Humphries (1977) and Mark-Lawson and

158

Witz (1986) theorise more rigorously the patriarchal gender relations which underlie this system of production.

9. This is the case in protestant industrialised countries, particularly Britain and North America, but cannot be generalised across Europe (for studies of various southern European countries see Carvacao, 1981; Garcia and Canoves, 1988; Strategaki, 1988).

10. The rise (and fall) of the concept of 'locality' and its widespread application in 'locality studies' in British geography in the 1980s is reviewed in Duncan, 1985 and Duncan and Savage, 1989. While some very good work has been inspired by this approach (e.g. Murgatroyd *et al.*, 1984; Cooke, 1989) much has failed to rise above the spatial fetishism of the old 'regional studies' which it superseded. For an example of its application to rural studies see Bradley and Lowe (1984).

11. As work in the United States, for example, makes clear, 'race' represents an important dimension of the social organisation of family farming (Jones, 1988). All the families and individuals involved in this research are white. This clearly does not preclude the significance of 'race' to the specific circumstances of this study. However the ideologies of nationalism and 'Englishness' associated with farming and expressed through the metaphors of 'breeding' and 'bloodstock' are not pursued here. I am grateful to Peter Jackson for bringing these themes to my attention.

2 Family Farming

1. The persistence of family farms has been an issue of particular significance to Marxist theorists, but comparable arguments and expectations can also be found in the work of Weber, and neo-classical economists (see Buttel and Newby, 1980; Marsden *et al.*, 1986a).

2. See for example: Lenin, 1946; Mouzelis, 1976; Ennew *et al.*, 1977; Vergopolos, 1978; Goodman and Redclift, 1981. For a useful reappraisal see the special issue of *Social Analysis* (vol. 20, 1986).

3. This problematic was shared with parallel non-Marxist political economy traditions. See for example: Stinchcombe, 1961; Barkley, 1976; Newby *et al.*, 1978; Ghorayshi, 1986. Prominent examples within Marxist political economy are: Goss *et al.*, 1980; Conway, 1981; Mooney, 1983.

4. Important contributions to Marxist rent theory include Massey and Catalano, 1978; Harvey, 1983; Ball *et al.*, 1985.

5. This topic has received considerably more attention in the United States than in the United Kingdom, but see Clutterbuck and Lang, 1981; Burns *et al.*, 1983; Ilbery and Healy, 1985; Barlow, 1988, and Marsden and Little, 1990.

6. Examples include salad and fruit crop production (e.g. Freidland *et al.*, 1981; de Janvry, 1980a; Thomas, 1985); and livestock production (Marsden and Symes, 1985). Recent developments in biotechnology suggest further, rapid transformations (Goodman *et al.*, 1987).

7. Terminological confusion abounds in attempts to specify the

distinctiveness of family, or household, enterprises, including: petty commodity production (PCP), simple commodity production (SCP), and domestic commodity production (DCP). Despite their loose usage in the literature, these terms are not strictly interchangeable, as there are important differences in the way in which they define their object of study. The concept DCP is employed here for reasons explained in Chapter 3. Scott (1986b) provides a useful discussion of these issues.

8. This is the result in large part of Friedmann's wider conception of capitalism as the 'generalised circulation of commodities, especially labour power' (1980, p. 160). A stronger basis for analysing the external relations of production of the family farm is the conception adopted in the 'constraints thesis' in terms of its essential relations of production. See Bernstein (1986) for a full discussion of this point.

9. Uneven development in agriculture, due to the substantial influence of land quality as a means of production, has been explored elsewhere in terms of its impact on differential patterns of agricultural development (see, for example, Fitzsimmons, 1986; Marsden *et al.*, 1987).

10. This research team was based at the Department of Geography, University College London. Directed by Professor Richard Munton (UCL) and Dr Terry Marsden (Southbank Polytechnic). It involved the author as research assistant on three related projects funded by the Economic and Social Research Council (ESRC) from November 1984 to October 1987. It also involved two other research assistants, Dr Jo Little and Mr Jack Eldon, for various stages/periods of the project.

11. Harré (1981) distinguishes taxonomic collectives, as the classification of groups on the basis of common morphological features of a phenomenon, from relational or structural groups, classified by their common causal properties, which hence have a potential for explaining change (see also Allen, 1983).

3 A Feminist Reconstruction

1. Critiques of structural–functionalist theories of the family are extensive (see, for example, Morgan, 1975; Harris, 1978; Thorne and Yalom, 1982, in the family sociology literature; and Beechey 1978; Barrett and McIntosh, 1982; and Caldwell, 1984, in the feminist and gender studies literature).

2. The family, like ideology and the state, was relegated to the level of superstructure in structuralist Marxist analyses, but received far less attention. Exceptions include Poster, 1978 and Kuhn, 1978. It came to receive most attention in terms of the Marxist/feminist 'domestic labour debate' (see, for example, Harrison, 1974; Gardiner, 1975). Molyneux (1979) and Kaluzynska (1980) provide thoroughgoing feminist critiques of the terms of this debate.

3. These questions relate to wider debates about gender and stratification (see, for example, Murgatroyd *et al.*, 1984; Crompton and Mann, 1986) and the 'naturalisation' of the household in historical research (such as Chaytor, 1980; O. Harris, 1982); in the modern 'third world' (such as Whitehead, 1981) and 'first world' (such as Gray, 1978).

4. This criticism lies at the heart of feminist critiques (see, for example, Molyneux, 1977; Mackintosh, 1979) of Marxist anthropological studies of household production systems in developing countries associated with the debate surrounding the 'articulation of modes of production' (Wolpe, 1980).

5. In her reply to Goodman and Redclift's criticisms of her work, for example, Friedmann (1986a) incorporates the term 'patriarchy' in the title of a paper. It receives limited attention in the text and is used to refer to the control of the father over the transfer of family property to male heirs in the process of succession.

6. Connell draws attention to the necessary extension of these arguments to the biological categories of male and female which are themselves not simply dimorphic or fixed (1987; ch. 2). See also Plumwood, 1989, for an examination of these issues beyond what is possible here.

7. There are a growing number of exceptions to the general monopoly of critical research on gender relations and sexuality from feminist perspectives. These include work from men's studies (Hearn, 1983; Brittan, 1988) and work informed by the politics of gay liberation (Weeks, 1985). One of the most exciting and accessible is that by Connell, 1987, who adopts a 'theory of practice' approach and retains women's oppression within contemporary gender relations as a central focus.

8. See the debate in *Antipode*, following Foord and Gregson's contribution (issues 18/3, 19/1, 19/2, 20/1). Most seriously, *gender relations* and *mode of production* are not equivalent objects of analysis, or theoretical concepts, as they suggest. Rather it is *mode of production* and *mode of reproduction* which are logically the autonomous objects of analyses implied by their method of abstraction and supported by their subsequent analysis. Class and gender represent the necessary social relations associated with these objects.

9. This is not an inherent shortcoming of realist methods (Sayer, 1984), but one arising from aspects of Foord and Gregson's application of it.

10. For example, Delphy's concept of a 'patriarchal mode of production' (1984), recently extended by Walby (1989), embodies this problem.

11. R. Williams (1983) advocated the term 'livelihood' instead of the traditional Marxist concern with production, as a means of shifting analytical and political weight away from the organisation of capital to the people's everyday economic activities and experiences.

12. This dualism is problematic, as a concept of 'social production' perpetuates the notion that household or subsistence production is 'non-social'. However this terminology is retained here in preference to Bryceson's (1983) alternative concepts of abstract and concrete labour, as these seem to add to, rather than reduce, the terminological confusion.

13. Marxist concepts of labour, along with neo-classical concepts (Roston, 1983), can in this way be said to be androcentric in the sense that they systematically devalue, or render invisible, major areas of women's work (see Holstrom, 1981).

14. Historical work (such as Bourdieu, 1976; Rapp *et al.*, 1979) and

anthropological research (such as Yanagisako, 1979; Moore, 1988) indicate the diverse forms taken by each of these component elements as well as their interrelations in specific times and places.

15. The argument is thus not that filial relations are unimportant in terms of family labour and property divisions, but that these have been emphasised at the expense of gender (Creighton, 1980; Symes and Appleton, 1986; Hutson, 1987) and are themselves structured by patriarchal gender relations (Qvortrup, 1985).

16. A total of 63 per cent of households in the United Kingdom can be classified in this category, including households composed of heterosexual couples with and without children (EOC, 1987).

17. This occurs, for example, through such practices as skill designation (see, for example, Phillips and Taylor, 1980; Game and Pringle, 1983; Cockburn, 1985).

18. See, for example, Pahl's tripartite division of work between the formal economy, underground economy and household or communal economy (1984, p. 118), and Sharpe's fourfold categorisation of irregular, formal, household and communal work (1988).

19. This fourfold division is not exhaustive. Most notably it does not include women's extensive voluntary off-farm labour, which cements local community (Bouquet, 1981; Stebbing, 1984) and wider kinship ties (di Leonardo, 1987).

4 Theory into Practice

1. The literature on feminist research methods is extensive (see particularly, Roberts, 1981; Oakley, 1981; EOC, 1986).

2. The argument here is not that of Dale Spender in *Man-made Language* (1980), who argues that established language is male, but that of Assiter (1983), who argues that it is in the use, or meaning, of language, rather than in its etymology or referents, that patriarchal gender relations are incorporated and reproduced. The distinction is an important one because, like Kaplan (1986), I would argue against recent versions of feminist critique which seek to construct a separate women's language/discourse.

3. Their footnote was itself a response to the insistence of the journal's editor that they specify the exact degree of bias arising from their method! (Marsden, personal communication).

4. This dissatisfaction with positivist methods in social science is wideranging. Significant elements include: the realist critique of positivist science (Keat and Urry, 1975; Sayer, 1984); the revitalisation of ethnographic methods (Knorr-Cetina and Cicourel, 1981; Eyles and Smith, 1988) and the extension of discourse analysis from literary and cultural studies (Hall *et al.*, 1980; Weedon, 1987).

5. Notable amongst this work are Bourdieu's (1977) concepts of 'habitus' and social 'dispositions'; Bhaskar's (1978) 'transformational model of the society/person relationship'; Giddens's (1979) 'structuration theory'; Berger and Luckmann's (1983) 'social construction of reality'; and Heller's (1984) concept of 'everyday sense making'.

6. On realist methods see Sayer, 1984; Massey and Meegan, 1985; Sarre, 1987. See also Gregson's attempt (1987) to draw out some of the methodological implications of Giddens's structuration theory.

7. It is worth noting that the farmer, the respondent in the project's base-line survey, was, in all but five cases in the sample of 185 farms, a man. This is close to the national figure of some 90 per cent mentioned in the introduction.

8. As project research assistant I carried out the 'farm survey' interviews in the MGB. Those in Dorset were carried out by Jo Little.

9. Fifty farms were discounted from the base-line sample for the survey of farm wives. These included: 19 cases of unmarried, divorced and widowed men; eight non-family businesses; five unmarried or widowed women farm principals; and 18 very elderly retired or ill couples who it was thought best not to trouble further. In the case of the men farmers without wives it should be noted that women housekeepers and other female kin, such as mothers and sisters, filled the domestic labour role of the absent 'wife' (see Gittins, 1985).

10. This response rate is comparable to that of Buchanan *et al.*, 1982, who conducted a postal survey of farm wives in the Reading area with a sample of 156 and a response rate of 60 per cent. They put this high response rate down to farm wives' 'latent desire to tell their story' (p. 23).

11. Hobby farms are defined as non-commercial agricultural concerns where the main source of livelihood for the occupying household is not derived from farming. The life-styles and standards of living of these households may be characterised as more representative of those associated with the 'stockbroker belt'.

12. Out of eight women initially approached to be involved in the case study exercise, seven women were recruited to participate in the exercise. One of these cases involved a 'mother and daughter', both farm wives, who wanted to be jointly involved in the project. However this case turned out not to work well within the wider framework, partly because the daughter's farm had not been included in the base-line farm survey, and partly because a shallower insight was gained about each of the two households/enterprises involved than in the cases focusing on just one family farm. This case was thus dropped from the final analysis.

13. The nearest comparisons to this method can be found in 'family sociology' (e.g. Laslett and Rappaport, 1975; Berk, 1980).

14. This sentiment is not peculiar to farm women but is also a feature of men's descriptions of their lives as farmers, although expressed in rather different ways and with different connotations.

5 Women's Work and Property

1. This refers to elderly relatives resident in the conjugal family household but excludes those living elsewhere who may be attended by farm wives.

2. The sole responsibility for domestic household labour falling on the

farm wife is in contrast to the historical position in which households incorporated female servants and kin who shared such responsibilities (see Bouquet, 1986; Chaytor, 1980).

3. While clearly age-related, the role of daughters as domestic workers is also influenced by the social relations of childhood, which are themselves actively gendered (Qvortrup, 1985) and class specific. For example, whether children go to private boarding schools or local state schools is likely to make a difference to the household gender division of labour, as well as to the construction of children's gender identities.

4. Such problems have received official recognition amongst development aid organisations on the basis of research findings which point consistently to the significance of the economic contribution of women to agrarian production (see Dixon-Mueller, 1985; Mies, 1986).

5. Women's field labour in horticultural production is more extensive, on a predominantly seasonal waged basis, but there are no such cases in these two study areas.

6. Comparing these results with those of other surveys of farm wives in England (Buchanan *et al.*, 1982) stresses not only the importance of the way women's labour contribution is assessed but also the kinds of farms which make up the sample.

7. Primary responsibility is defined as providing the staple labour and managing the production process.

8. The more common national pattern is to give up employment on the birth of a first child (EOC, 1987).

9. There is an extensive literature on women's off-farm employment, particularly in the United States centred on the question of whether or not it is a 'good thing' for the family farm. Traditionally it had been argued that wives' off-farm employment weakened the integrity of the family labour process (e.g. Sweet, 1972). More recently, analyses have sought to show such work to be essential to household income (e.g. Buttel and Gillespie, 1984 (United States); Gasson, 1984 (western Europe), or to promote marital stability by 'providing an outlet for women's creative energies' (Acock and Deseran, 1986, p. 317).

10. A total of 70 per cent of farmland in England and Wales is owner-occupied. Table 4.1 shows the tenure pattern in the two study areas.

11. Five women inherited land from their fathers, four from their husbands, three from both their father and their husband, and one from her mother (three did not specify the source).

12. Three women inherited capital assets from their fathers, two from their husbands, two from both their fathers and husbands, and one from her mother (three did not identify the source).

13. Long-term, or strategic decisions refer to major investment decisions, changes in business organisation etc.; day-to-day decisions refer to those connected with the daily management of the farm.

14. These findings contradict those of Buchanan *et al.* (1982) for the Reading area, who recorded between 62 and 68 per cent of women regularly involved in farm decision-making. This may be partly

explained by the relatively high non-response rate to this question in their survey (20 per cent) as against only one missing case here.

15. The designation of farm wives as 'self-employed' for accounting and taxation purposes appears to be rising, although, unsurprisingly, it is difficult to gauge reliably. However, in the case study interviews, it was widely discussed as a mechanism for reducing liabilities.

16. Similar findings (see Hunt, 1978; Mackenzie and Rose, 1983) have been reported showing that men's unemployment in urban contexts leads to an enforcement rather than a relaxation of the gender division of domestic household labour.

6 'Being the Farmer's Wife'

1. This term derives from women's own description of their experience as 'reserve labour' on the farm, rather than from the Marxist–feminist concept of the 'reserve army of labour' (Beechey, 1977; Breughel, 1979; Anthias, 1980).

2. This was in marked contrast to my experience of interviewing men as 'farmers'.

3. Compare with Ruth Gasson's threefold distinction between housewives, working women and farming partners (1981).

4. Clearly this is not to argue that within the terms of their known lifeworld women (and men) do not rationalise their actions, nor that concrete knowledge cannot be informed by rational abstractions (see the arguments in Bourdieu, 1977, about the links between ideology, practice and scientific discourse).

5. The principle of using research subjects' own definition of their life experience is elaborated by Wallman (1984, p. 15).

7 The Domestic Political Economy of Six Family Farms

1. The value of comparative analysis implicit in the paired case study strategy is well established in anthropology and spelt out in Wallman (1984).

2. Followers are young stock, usually born to cows in the dairy herd. The males are kept until between 12 and 18 months old, when they are sold off for beef, while the females may be taken on as additional or replacement dairy stock when they mature, or are sold out.

3. Contract labour describes labour relations where the worker is technically self-employed and hires him/herself out at an hourly or daily rate, sometimes with specialist machinery, to do a particular job. It is commonly a means by which farmers earn extra income, but in some cases is a full-time form of employment for the person concerned. For the employing farmer it avoids keeping a worker on the whole year round, thereby escaping some of the statutory costs of employing full-time workers, and/or the capital cost of maintaining or replacing specialist machinery.

4. There is a notable gender division of labour amongst the hired

workforce, too. It is men who have the best pay and conditions of work as full-time workers and women who make up the part-time and, particularly, the seasonal and casual workforce, with poorer pay and conditions.

5. This figure is not returned as profit, which is much lower through various tax and accounting practices.

6. Julie Church's father was seriously ill at the end of the interview period. The possibility of her inheriting money or assets from him as an independent resource base has important implications for the future of her dried-flower business and the conjugal farm.

7. Mr Brown is unusual in that he is not from a farming background, but rather trained at Vickers as a fitter. The smallholding route thus represented his only means into farming.

8. Under smallholdings legislation, the land-owning council can impose a number of restrictions on the operations of smallholders. In this case, the council concerned is operating a policy of selling off its smallholding land to raise revenue; it is therefore discouraging the more marginal producers by forcing them to sell only what they can grow on their own land.

9. Mr Watson's fears were inadvertently substantiated when he, and his wider family, came up in a conversation with another of the case study families in the area. Judgement was passed on his inability to earn a living at farming.

8 Conclusions

1. I am not here supporting the position of Delphy (1984), and more recently Walby (1986), who term the exploitation of women's domestic labour a patriarchal mode of production. Such a position assumes a very ethnocentric generalisation of family/household form and the conditions of women's domestic work. Moreover the idea of a patriarchal mode of production represents a misapplication of this term, similar to the search in early theories of agrarian change for a peasant, or simple commodity, mode of production.

2. In particular, the ethnographic method of cumulative interviewing adopted here is best suited to research on particular institutions, such as family households or workplaces. Participant observation methods would be more appropriate to the study of either occupational or local community ideologies.

3. This is in contrast to most of the feminist analyses of patriarchal work relations (see for example, Cockburn, 1985; Walby, 1986) which stress the significance of collective experience and action, such as in unions, through which men and women contest their work conditions.

4. These organisations were formed, and continue primarily, to address other farming issues. However both are now involved in the Women's Committee of COPA, the European Farming Union which brings together women from all member states of the EEC to campaign and lobby for women's status and rights in farming.

5. In an interview on Radio 4's 'Woman's Hour' as recently as June 1989, a senior representative of the National Farmers' Union was still claiming that the directive had little relevance in Britain, because women did very little manual work on farms as a result of the larger size of farms here compared with elsewhere in the EEC. However, as a result of the publicity surrounding the directive, they have since organised a conference on the issue.

Bibliography

ACOCK, A. and F. DESERAN (1986) 'Off-farm employment by women and marital stability', *Rural Sociology*, 51, pp. 314–27.

AGRICULTURE ECONOMIC DEVELOPMENT COMMITTEE (1972) *Agricultural Manpower in England and Wales* (London: National Economic Development Office, HMSO).

ALBERT, C., M. BERLAN, J. CANIOU and M. PERROT (1987) *Celles de la Terre. Agricultrice: l'Invention Politique d'un Métier* (Paris: Editions EHESS).

ALLATT, P., T. KEIL, A. BRYMAN and B. BYTHEWAY (eds) (1987) *Women and the Life Cycle: Transitions and Turning Points* (London: Macmillan).

ALLEN, J. (1983), 'Property relations and landlordism – a realist approach', *Society and Space*, 1, pp. 191–203.

ANDRE, J. (1985) 'Power, oppression and gender', *Social Theory and Practice*, 11/1, pp. 107–22.

ANTHIAS, F. (1980) 'Women and the reserve army of labour: a critique of Veronica Beechey', *Capital and Class*, 10, pp. 50–63.

ARDENER, E. (1978) Gender and language, in S. Ardener (ed.), *Defining Females: the Nature of Women in Society* (London: Croom Helm).

ASSITER, A. (1983) 'Did man make language?' *Radical Philosophy*, 34, pp. 25–9.

AYIM, M. and B. HOUSTON (1985) 'The epistemology of gender identity, *Social Theory and Practice*, 11/1, pp. 25–60.

BALL, M., (1979) 'On Marx's theory of agricultural rent: a reply to Ben Fine', *Economy and Society*, 19, pp. 304–26.

BALL, M., B. BENTIVEGNA, M. EDWARDS and M. FOLIN (eds) (1985) *Land Rent, Housing and Urban Planning* (London: Croom Helm).

BANAJII, J. (1976) 'Kautsky's agrarian question', *Economy and Society*, 5, pp. 2–49.

BARKER, D. and S. ALLEN (eds) (1976) *Dependence and Exploitation in Work and Marriage* (London: Longman).

BARKLEY, P. (1976) 'A contemporary political economy of family farming', *American Journal of Agricultural Economics*, 58, pp. 812–19.

BARLOW, J. (1988) 'A note on biotechnology and the food production chain. Some social and spatial implications of changing production technology', *International Journal of Urban and Regional Research*, 12, pp. 229–46.

BARRETT, M. (1980) *Women's Oppression Today, Problems in Marxist Feminist Analysis* (London: Verso).

BARRETT, M. and M. McINTOSH (1982) *The Anti-social Family* (London: Verso).

BARTELLS, A. (1982) 'Biological sex differences and sex stereotyping', in *The Changing Experience of Women* (Open University) pp. 254–66.

BARTHEZ, A. (1983) *Famille, Travail et Agriculture* (Paris: Economica).

BARTHEZ, A. (1985) 'Les agricultrices, travailleuses à part entière', in J. Poteh (ed.), *L'état de la France et de ses Habitants* (Paris: La découverte).

BAUWENS, A. and T. LOEFFEN (1983) 'The changing economic and social position of the farmer's wife in the Netherlands', paper to the 12th congress for Rural Sociology, July, Budapest.

BECHHOFER, F. and B. ELLIOTT (eds) (1981) *The Petite Bourgeoisie* (London: Macmillan).

BEECHEY, V. (1977) 'Some notes on female wage labour in the capitalist mode of production', *Capital and Class*, 3, pp. 45–66.

BEECHEY, V. (1978) 'Women and production: a critical analysis of some sociological theories of women's work', in A. Kuhn and A. Wolpe (eds), *Feminism and Materialism* (London: Routledge & Kegan Paul), pp. 155–96.

BENERIA, L. (1979) 'Reproduction, production and the sexual division of labour', *Cambridge Journal of Economics*, 3, pp. 203–25.

BENERIA, L. (1982a) 'Accounting for women's labour', *Women and Development*, pp. 119–48.

BENERIA L. (ed.) (1982b) *Women and Development. The Sexual Division of Labour in Rural Societies* (New York: Praeger).

BENNHOLT-THOMPSON, V. (1982) 'Subsistence production and extended reproduction. A contribution to the discussion of modes of production', *Journal of Peasant Studies*, 9, pp. 241–54.

BENTON, T. (1984) *The Rise and Fall of Structural Marxism. Althusser and his Influence* (London: Macmillan).

BENVENUTI, B. (1985) 'On the dualism between sociology and rural sociology: some hints for the case of modernisation', *Sociologia Ruralis*, 25, pp. 214–30.

BERGER, P. and T. LUCKMANN (1983) *The Social Construction of Reality* (Harmondsworth: Penguin).

BERK, S. FENSTERMAKER (ed.) (1980) *Women and Household Labour* (London: Sage).

BERK S. F. and SHIH A. (1980) 'Contributions to household labour. Comparing wives and husbands views', in S. F. Berk (ed.) *Women and Household Labour*, pp. 191–227.

BERNADES, J. (1985) 'Do we really know what the family is?' in P. Close and R. Collins (eds), *Family and Economy in Modern Society* (London: Macmillan) pp. 192–211.

BERNSTEIN, H. (1986) 'Capitalism and petty commodity production', *Social Analysis*, 20, pp. 9–26.

BHASKAR, R. (1978) 'On the possibility of social scientific knowledge and the limits of naturalism', *Journal for the Theory of Social Behaviour*, 8/1, pp. 1–28.

BLAND, L., R. HARRISON, F. MONT and C. WEEDON (1978) 'Relations of reproduction: approaches through anthropology in CCCS', *Women Take Issue*, pp. 155–75.

BOSERUP, E. (1970) *Women's Role in Economic Development* (New York: St Martin's Press).

BOULDING, E. (1980) 'The labor of US farm women: a knowledge gap', *Sociology of Work and Occupations*, 7/3, pp. 261–90.

BOUQUET, M. (1981) 'The Sexual Division of Labour: the Farm House-hold in a Devon Parish', PhD thesis, Cambridge.

BOUQUET, M. (1982) 'Production and reproduction of family farms in south west England', *Sociologia Ruralis*, 22/3, pp. 227–44.

BOUQUET, M. (1984a) 'Women's work in rural south west England', in N. Long (ed.), *Family and Work in Rural Societies* (London: Tavistock), pp. 142–59.

BOUQUET, M. (1984b) 'The differential integration of the rural family', *Sociologia Ruralis*, 24/1, pp. 65–75.

BOUQUET, M. (1986) *Family, Servants and Visitors* (Norwich: Geobooks).

BOURDIEU, P. (1976) 'Marriage strategies as strategies of social reproduction', in R. Forster and O. Ranum (eds), *Family and Society: Selections from the Annales* (Baltimore: Johns Hopkins University Press), pp. 117–44.

BOURDIEU, P. (1977) *Outline of a Theory of Practice* (Cambridge: Cambridge University Press).

BOWLBY, S., J. FOORD, J. LEWIS and L. McDOWELL (1986) 'The place of gender in locality studies', *Area*, 18/4, pp. 327–31.

BRADBY, C. (1975) 'The destruction of the natural economy', *Economy and Society*, 4, pp. 125–61.

BRADBY, B. (1977) 'Male rationality in economics. A critique of Godelier on salt money. The non-valorisation of women's labour', *Critique of Anthropology*, 3/9, pp. 131–8.

BRADBY, B. (1982) 'The remystification of value', *Capital and Class*, 17, pp. 114–33.

BRADLEY, T. (1981) 'Capitalism and the countryside: rural sociology as political economy', *International Journal of Urban and Regional Research*, 5, pp. 581–7.

BRADLEY, T. and P. LOWE (eds) (1984) *Locality and Rurality: Economy and Society in Rural Regions* (Norwich: Geobooks).

BRASS, T. (1986) 'The elementary structures of kinship: unfree relations and the production of commodities', *Social Analysis*, 20, pp. 56–68.

BREUGHEL, I. (1979) 'Women as a reserve army of labour: a note on recent British experience', *Feminist Review* 3, pp. 12–23.

BRIDENTHAL, R. (1979) 'Family and Reproduction', in R. Rapp, E. Ross and R. Bridenthal, 'Examining Family History', *Feminist Studies*, 5/1.

BRITTAN, A. (1988) *Masculinity* (Cambridge: Polity Press).

BROMLEY, R. and C. GERRY (eds) (1979) *Casual Work and Poverty in Third World Cities* (Chichester: John Wiley).

BRYCESON, D. F. (1983) 'Use values, the law of value and the analysis of non-capitalist production', *Capital and Class*, 20, pp. 29–63.

BUCHANAN, W. I., A. ERRINGTON and A. GILES (1982) *The Farmer's Wife: her Role in the Management of the Business* (University of Reading: Farm Management Unit) study no. 2.

BURAWOY, M. (1979) *Manufacturing Consent* (Chicago: University of Chicago Press).

BURAWOY, M. (1985) *The Politics of Production* (London: Verso).

BURGESS, J., M. LIMB and C. HARRISON (1988) 'Exploring environmental values through the medium of small groups: 1 Theory and practice', *Environment and Planning A*, 20, pp. 309–26.

BURMAN, S. (ed.) (1979) *Fit Work for Women* (London: Croom Helm).
BURNS, J., J. MCINERNEY, and A. SWINBANK (eds) (1983) *The Food Industry. Economics and Politics* (London: Heineman).
BURRELL, A., B. HILL and J. MEDLAND (1984) *A Statistical Handbook on UK Agriculture* (London: Macmillan).
BUTTEL, F. (1982) 'The political economy of agriculture in advanced industrial societies', *Current Perspectives in Social Theory*, 3, pp. 27–55.
BUTTEL, F. and H. NEWBY (eds) (1980) *The Rural Sociology of Advanced Societies* (Montclair, NJ: Allanheld, Osmun and Co.)
BUTTEL, F. and G. W. GILLESPIE (1984) 'The sexual division of labour: an exploratory study of the structure of on-farm and off-farm labour allocation among farm men and women', *Rural Sociology*, 49/2, pp. 183–209.
CALDWELL, L. (1984) Feminism and the family: a review article, *Feminist Review*, 16, pp. 88–97.
CALLAN, H. and S. ARDENER (1984) *The Incorporated Wife* (London: Croom Helm).
CARVACAO, C. A. (1981) 'A mulher na agricultura Portuguesa', Estudios de Geografica Humana e Regional, University of Lisbon, mimeo.
CENTRE FOR CONTEMPORARY CULTURAL STUDIES (1978) *Women Take Issue* (London: Hutchinson).
CHAYANOV V. I. (1966) (1st edn 1925) in D. Thorner, R. Smith and B. Kerblay (eds), *The Theory of Peasant Society*. (London: Irwin).
CHAYTOR, M. (1980) 'Household and kinship: Ryton in the late 16th and early 17th centuries', *History Workshop Journal* 9–10, pp. 25–60.
CHEVALIER, J. (1983) 'There's nothing simple about simple commodity production', *Journal of Peasant Studies*, 10/4, pp. 153–86.
CLOSE, P. and R. COLLINS (eds) (1985) *Family and Economy in Modern Society* (London: Macmillan).
CLUTTERBUCK, C. and T. LANG (1982) *More Than We Can Chew: The Crazy World of Food and Farming* (London: Pluto).
COCHRANE, W. (1958) *Farm Prices: Myth and Reality* (Minneapolis: University of Minnesota Press).
COCKBURN, C. (1985) *Male Dominance. Women and Men and Technical Know-how* (London: Pluto).
COCKBURN, C. (1986) 'The relations of technology. What implications for theories of sex and class?' in R. Crompton and M. Mann (eds), *Gender and Stratification* (Cambridge: Polity Press), pp. 74–85.
COLLINS, R. (1985) 'Horses for courses': Ideology and the division of domestic labour', in P. Close and R. Collins, *Family and Economy in Modern Society* (London: Macmillan) pp. 64–83.
COMER, L. (1982) 'Monogamy, marriage and economic dependence', *The Changing Experience of Women*, pp. 178–89.
CONNELL R. (1983) 'The concept of role and what to do with it'. In R. Connell, *Which Way is Up?* pp. 189–207.
CONNELL R. (1985) 'Theorising gender', *Sociology*, 19/2, pp. 260–72.
CONNELL R. (1987) *Gender and Power* (Cambridge: Polity Press).
CONWAY, J. (1981) 'Agrarian petit-bourgeois responses to capitalist industrialisation: the case of Canada', in F. Bechhofer and B. Elliott, *The Petite Bourgeoisie* (London: Macmillan) pp. 1–37.

COOKE, P. (ed.) (1989) *Localities* (London: Unwin Hyman).

COULSON, M., B. MAGAS and H. WAINWRIGHT (1975) 'The housewife and her labour under capitalism', *New Left Review*, 89.

COWARD, R. (1983) *Patriarchal Precedents* (London: Routledge & Kegan Paul).

COX, G., P. LOWE and M. WINTER (eds) (1986) *Agriculture: People and Policies* (London: Allen & Unwin).

CREIGHTON, C. (1980) 'Family, property and relations of production in western Europe', *Economy and Society*, 9/2, pp. 129–67.

CROMPTON, R. and M. MANN (eds) (1986) *Gender and Stratification* (Cambridge: Polity Press).

CROW, G. (1985) 'The farm and the village: representations of post-war family farming in British community studies', mimeo.

CROW, G. (1989) 'The use of the concept "strategy" in recent sociological literature', *Sociology*, 23/1, pp. 1–24.

CURTIN, C. (1986) 'The peasant family farm and commoditisation in the west of Ireland', in Long *et al.*, *The Commoditisation Debate*, pp. 58–76.

DAVIDOFF, L. (1986) 'The role of gender in the first industrial nation. Agriculture in England 1780–1850', in Crompton R. and M. Mann, *Gender and Stratification* (Cambridge: Polity Press) pp. 190–213.

DAVIDOFF, L. and C. HALL (1987) *Family Fortunes. Men and Women in the English Middle Class 1780–1850* (London: Hutchinson).

DAVIS, J. E. (1980) 'Capitalist agricultural development and the exploitation of the propertied labourer', in F. Buttel and H. Newby, *Rural Sociology in Advanced Societies* (Montclair, NJ: Allanheld, Osmun and Co.).

DELPHY, C. (1984) *Close to home* (London: Hutchinson).

DELPHY, C. and D. LEONARD (1986) 'Class analysis, gender analysis and the family', in R. Crompton and M. Mann, *Gender and Stratification* (Cambridge: Polity Press) pp. 57–75.

DIXON-MUELLER, R. (1985) *Women's Work in Third World Agriculture* (Geneva: International Labour Office) Women, Work and Development, series no. 9.

DONZELOT, J. (1979) *The Policing of Families: Welfare Versus the State* (London: Hutchinson).

DUNCAN, S. (1985) *What is Locality?* University of Sussex, Sussex working paper no. 51.

DUNCAN, S. and M. SAVAGE (1989) 'Space, scale and locality', *Antipode*, 21/3, pp. 179–206.

EDHOLM, E. (1982) 'The unnatural family', *The Changing Experience of Women*, pp. 166–77.

EDHOLM, E., O. HARRIS and K. YOUNG (1977) 'Conceptualising women', *Critique of Anthropology*, 3/9–10, pp. 101–30.

EISENSTEIN, H. (1982) 'Which way out of the impasse? The politics of feminist theory in the 1980s', *Thesis Eleven*, 5/6, pp. 259–70.

ENNEW, J. P. HIRST and K. TRIBE (1977) 'Peasantry as an economic category', *Journal of Peasant Studies*, 4, pp. 295–322.

EQUAL OPPORTUNITIES COMMISSION (1986) *Methodological Issues in Gender Research*, EOC Research Bulletin no. 10.

EQUAL OPPORTUNITIES COMMISSION (1987) *Women and Men in the United Kingdom* (London: HMSO).

ERRINGTON, A. (1983) 'The farmer's wife: her role in the farm business', in R. B. Tranter (ed.), *Strategies For Family Worked Farms in the UK* (Reading: Centre for Agricultural Strategy) paper no. 15, pp. 223–8.

ERRINGTON, A. (1985) 'Sampling frames for farm surveys in the UK: some alternatives', *Journal of Agricultural Economics*, 36/2, pp. 251–8.

ERRINGTON, A. (1987) 'Rural employment trends and issues in market industrialised countries. Report for the International Labour Office', mimeo.

ERRINGTON, A., (ed) (1986) *The Farm as a Family Business. An annotated bibliography* (Stoneleigh: Agricultural Manpower Society).

EYLES, J. and D. SMITH (eds) (1988) *Qualitative Methods in Human Geography* (Cambridge: Polity Press).

FINCH, J. (1983) *Married to the Job: Wives Incorporation in Men's Work* (London: Allen & Unwin).

FINE, B. (1978) 'On Marx's theory of agricultural rent', *Economy and Society*, 8, pp. 241–78.

FITZSIMMONS, M. (1986) 'The new industrial agriculture', *Economic Geography*, 62, pp. 334–53.

FLAX, J. (1982) 'The family in contemporary feminist thought', in J. Elshtain (ed), *The Family in Political Thought* (Brighton: Harvester Press) pp. 223–53.

FLORA, C. BUTLER (1981) 'Farm women, farming systems and agricultural structure', *The Rural Sociologist*, 1, pp. 383–6.

FOLBRE, N. (1982) 'Exploitation comes home: a critique of the Marxian theory of family labour', *Cambridge Journal of Economics*, 6, pp. 317–29.

FOORD, J. and N. GREGSON (1986) 'Patriarchy: towards a reconceptualisation', *Antipode*, 18/2, pp. 186–211.

FOORD, J., L. McDOWELL and S. BOWLBY (1986) *For 'love' not money: gender relations in local areas*, CURDS discussion paper no. 76, Newcastle University.

FRIEDLAND, W. H. (1982) 'The end of rural society and the future of rural sociology', *Rural Sociology*, 47, pp. 589–608.

FRIEDLAND, W. H., M. FURNARI and E. PUGLIESE (1981) 'The labour process and agriculture', paper given at the Conference on the Labour Process, University of Santa Cruz, California.

FRIEDMANN, H. (1978a) 'World market, state and family farm: social bases of household production in the era of wage-labour', *Comparative Studies in Society and History*, 20, pp. 545–86.

FRIEDMANN, H. (1978b) 'Simple commodity production and wage labour in the American plains', *Journal of Peasant Studies*, 6, pp. 70–100.

FRIEDMANN, H. (1980) 'Household production and the national economy: concepts for the analysis of agrarian formations', *Journal of Peasant Studies*, 7, pp. 154–84.

FRIEDMANN, H. (1981) 'The family farm in advanced capitalism: outline of a theory of simple commodity production', paper presented at the American Sociological Association, Toronto.

FRIEDMANN, H. (1986a) 'Property and patriarchy: a reply to Goodman and Redclift', *Sociologia Ruralis*, 26, pp. 186–93.

FRIEDMANN, H. (1986b) 'Family enterprises in agriculture: structural limits and political possibilities', in G. Cox *et al.*, (eds) *Agriculture: People and Policies* (London: Allen & Unwin) pp. 41–60.

FRIEDMANN, H. (1986c) 'Patriarchal commodity production', *Social Analysis*, 20, pp. 47–55.

FRIEDMANN, H. (1986d) 'Postscript', *Labour, Capital and Society*, 6, pp. 117–26.

GAMARNIKOW, E., D. MORGAN, J. PURVIS and D. TAYLORSON (eds) (1983) *The Public and the Private* (London: Heinemann).

GAME, A. and A. PRINGLE (1983) *Gender at Work* (London: Pluto).

GARCIA, M. and G. CANOVES (1988) 'The role of women in the family farm: the case of Catalonia', *Sociologia Ruralis*, 28/4, pp. 263–70.

GARDINER, J. (1975) 'The political economy of female labour in capitalist society', *New Left Review*, 88, pp. 47–58.

GARNSEY, E. (1981) 'The rediscovery of the division of labour', *Theory and Society*, 10/3, pp. 337–58.

GASSON, R. (1980) 'A questionnaire survey of farm women', *Farmer's Weekly*, various issues.

GASSON, R. (1981) 'Roles of women on farms', *Journal of Agricultural Economics*, 32/1, pp. 11–20.

GASSON, R. (1984) 'Farm women in Europe: their need for off-farm employment', *Sociologia Ruralis*, 24/3–4, pp. 222–7.

GASSON, R. (1986) 'Review of Bouquet's "Family, Servants and Visitors",' *Journal of Agricultural Economics*, 37, pp. 423–4.

GASSON, R. (1987) 'The careers of farmers' daughters', *Farm Management*, 6/7, pp. 309–17.

GASSON, R. (1989) 'Farm work by farmer's wives', *Farm Business Unit Occasional paper no. 15*, Wye College.

GASSON, R., G. CROW, A. ERRINGTON, J. HUTSON, T. MARSDEN and M. WINTER (1988) 'The farm as a family business. A review', *Journal of Agricultural Economics*, 39/1, pp. 1–41.

GEERTZ, C. (1973) *The Interpretation of Culture* (New York: Basic Books).

GEISLER, C. C., W. F. WATERS and E. L. EADE (1985) 'The changing structure of female land ownership 1946–1978', *Rural Sociology*, 50/1, pp. 74–87.

GHORAYSHI, P. (1986) 'The identification of capitalist farms. Theoretical and methodological considerations', *Sociologia Ruralis*, xxvi/2, pp. 146–69.

GHORAYSHI, P. (1989) 'The indispensable nature of wives' work for the farm family enterprise', *Canadian Review of Sociology and Anthropology*, 26(4), pp. 571–95.

GIDDENS, A. (1979) *Central Problems in Social Theory* (London: Macmillan).

GITTINS, D. (1985) *The Family in Question. Changing Households and Familiar Ideologies* (London: Macmillan).

GOODMAN, D. and M. REDCLIFT (1981) *From Peasant to Proletarian* (Oxford: Basil Blackwell).

GOODMAN, D. and M. REDCLIFT (1985) 'Capitalism, petty commodity production and the farm enterprise', *Sociologia Ruralis*, 25, pp. 231–47.

GOODMAN, D. and M. REDCLIFT (1988) 'Problems in analysing the agrarian transition in Europe', *Comparative Studies in Society and History*, 30/4, pp. 784–91.

GOODMAN, D., and M. REDCLIFT (eds) (1989) *The International Farm Crisis* (London: Macmillan).

GOODMAN, D., B. SORJ and J. WILKINSON (1987) *From Farming to Bio-technology: the Industrialisation of Agriculture* (Oxford: Basil Blackwell).

GOSS, K., R. RODEFIELD and F. BUTTEL (1980) 'The political economy of class structure in US agriculture: a theoretical outline', in Buttel F. and H. Newby, *The Rural Sociology of Advanced Societies* (Montclair, NJ: Allanheld, Osman and Co.) pp. 83–135.

GRAHAM, H. (1983) 'Do her answers fit his questions? Women and the survey method', in Gamarnikow *et al.*, *The Public and the Private* (London: Heinemann) pp. 132–46.

GRAY, A. (1978) 'The working-class family as an economic unit', in C. Harris (ed), *The Sociology of the Family*, pp. 186–213.

GREGORY, D. and J. URRY (eds) (1985) *Social Processes and spatial structures* (London: Macmillan).

GREGSON, N. (1987) 'Structuration theory: some thoughts on the possibilities for empirical research', *Society and Space*, 5, pp. 73–91.

GREGSON, N. and J. FOORD (1987) 'Patriarchy: a reply to critics', *Antipode*, 19/3, pp. 371–5.

GRIGG, D. (1987) 'Farm size in England and Wales from early Victorian times to the present', *Agricultural History Review*, 35, pp. 179–89.

HALL, S., D. HOBSON, A. LOWE and P. WILLIS (1980) *Culture, Media, Language* (London: Hutchinson).

HANEY, W. G. and J. B. KNOWLES (eds) (1988) *Women and Farming. Changing Roles, Changing Structures* (Boulder, Col: Westview Press).

HARRÉ, R. (1981) 'Philosophical aspects of the macro–micro problem', in K. Knorr-Cetina and A. V. Cicourel, *Advances in Social Theory and Methodology* (London: Routledge & Kegan Paul) pp. 139–60.

HARRIS, C. (1978) *The Sociology of the Family*, Sociological Review Monograph no. 28.

HARRIS, C. (1983) *The Family and Industrial Society* (London: Allen & Unwin).

HARRIS, O. (1981) 'Households as natural units', in K. Young *et al.*, *Of Marriage and the Market*, (London: Routledge & Kegan Paul) pp. 136–56.

HARRIS, O. (1982) 'Households and their boundaries', *History Workshop Journal*, 13/14, pp. 142–52.

HARRISON, A. (1975) *Farmers and farm businesses in England* (Reading: Dept. of Agricultural Economics and Management) miscellaneous study no. 62.

HARRISON, A. (1982) *Factors influencing the ownership, tenancy, mobility and use of farmland in the member states of the European Community*, Commission of the European Community, official publications no. 86.

HARRISON, A. (1983) 'Family farm policies in the European Community: are they appropriate for the UK?' in R. B. Tranter (ed.), *Strategies for family-worked farms in the UK* (Reading: Centre for Agricultural Strategy) paper no. 15, pp. 58–67.

HARRISON, J. (1974) 'The political economy of housework', *Bulletin of the Conference of Socialist Economists*, 3, pp. 35–52.

HART, N. (1989) 'Gender and the rise and fall of class politics', *New Left Review*, 175, pp. 19–47.

HARTMANN, H. (1979) 'The unhappy marriage of Marxism and Feminism: towards a more progressive Union', *Capital and Class*, 8, pp. 1–33.

HARVEY, D. (1983) *The Limits to Capital* (Oxford: Basil Blackwell).

HEARN, J. (1983) *Male Oppression* (Brighton: Harvester Press).

HEDLEY, M. (1981) 'Relations of production in the "family farm": Canadian Prairies', *Journal of Peasant Studies*, 9, pp. 71–85.

HELLER, A. (1984) *Everyday Life* (London: Routledge & Kegan Paul).

HIMMELWEIT, S. and S. MOHUN (1977) 'Domestic labour and capital', *Cambridge Journal of Economics*, 1, pp. 15–31.

HIRSCHON, R. (ed.) (1984) *Women and Property, Women as Property* (London: Croom Helm).

HOLSTROM, N. (1981) '"Women's work", the family and capitalism', *Science and Society*, 45, pp. 186–211.

HUMPHRIES, J. (1977) 'Class struggle and the persistence of the working-class family', *Cambridge Journal of Social Policy*, 7/3, pp. 241–58.

HUNT, P. (1978) 'Cash transactions and household tasks. Domestic behaviour in relation to industrial employment', *Sociological Review*, 26/2, pp. 555–71.

HUSSEIN, A. and K. TRIBE (1983) *Marxism and the Agrarian Question* (London: Macmillan).

HUTSON, J. (1987) 'Fathers and sons: family farms, family businesses and the farming industry', *Sociology*, 21, pp. 215–31.

ILBERY, B. and M. HEALY (eds) (1985) *The Industrialisation of the Countryside* (Norwich: Geobooks).

JANVRY, A. DE (1980a) 'Mechanisation in California agriculture: the case of canning tomatoes', Berkeley, *Dept. of Agriculture and Resource Economics paper*.

JANVRY, A. DE (1980b) 'Social differentiation in agriculture and the exploitation of neo-populism, in F. Buttel and H. Newby (eds), *The Rural Sociology of Advanced Societies* (Montclair, NJ: Allanheld, Osman and Co.).

JONES, C. A. and R. A. ROSENFELD (1981) *American Farm Women. Findings from a National Survey* (Chicago: University of Chicago).

JONES, J. (1988) 'Tore up and a-movin': perspectives on the work of black and poor white women in the rural South', in W. Haney and J. Knowles (eds), *Women and Farming* (Boulder, Col.: Westview Press), pp. 15–34.

KAHN, J. (1982) 'From peasants to petty commodity producers in Southeast Asia', *Bulletin of Concerned Asian Scholars*, 14, pp. 3–15.

KALUZYNSKA, E. (1980) 'Wiping the floor with theory: a survey of writings on housework', *Feminist Review*, 6, pp. 27–54.

KAPLAN, C. (1986) *Seachanges, Culture and Feminism* (London: Verso).

KAZI, H., S. LEES, and H. S. HIRZA (1986) 'Feedback: Feminism and Racism', *Feminist Review*, 22, pp. 87–105.

KEAT, R. and J. URRY (1975) *Social Theory as Science* (London: Routledge & Kegan Paul).

KLOPPENBURG J. (1988) *The First Seed* (Cambridge: Cambridge University Press).

KLOPPENBURG, J. and M. KENNEY (1984) 'Biotechnology and the restructuring of agriculture', *Insurgent Sociologist*, 12, pp. 3–17.

KNORR-CETINA, K. (1981) 'The micro-sociological challenge to macro-sociology: towards a reconstruction of social theory and methodology', in K. Knorr-Cetina and A. V. Cicourel (eds), *Advances in Social Theory and Methodology* (London: Routledge & Kegan Paul) pp. 1–48.

KNORR-CETINA, K. and A. V. CICOUREL (eds) (1981) *Advances in Social Theory and Methodology. Towards an Integration of Micro and Macro Sociologies* (London: Routledge & Kegan Paul).

KUHN, A. (1978) 'Structures of patriarchy and capital in the family', in A. Kuhn and A. Wolpe, *Feminism and Materialism* (London: Routledge & Kegan Paul) pp. 42–67.

KUHN, A. and A. WOLPE (eds) (1978) *Feminism and Materialism* (London: Routledge & Kegan Paul).

LAGRAVE, R. (1983) 'Bilan critique des recherches sur les agricultrices en France', *Etudes Rurales*, 92, pp. 9–40.

LASLETT, B. and R. RAPPAPORT (1975) 'Collaborative interviewing and interactive research', *Journal of Marriage and the Family*, 37, pp. 968–77.

LAWRENCE, R. J. (1982) 'Domestic space and society: a cross-cultural study', *Comparative Studies in Society and History*, 24, pp. 104–29.

LENIN, V. I. (1946) *Capitalism and Agriculture* (New York: International Publishers).

LEONARDO, M. DI (1987) 'The female world of cards and holidays; women, families and the work of kinship', *Signs*, 12/3, pp. 440–53.

LITTLEJOHN, J. (1963) *Westrigg* (London: Routledge & Kegan Paul).

LONG, N. (ed.) (1984) *Family and Work in Rural Societies: Perspectives on non-wage Labour* (London: Tavistock).

LONG, N. (1986) 'Commoditisation: thesis and anti-thesis', in N. Long *et al.* (eds), *The Commoditisation Debate*, pp. 8–23.

LONG, N., J. VAN DER PLOEG, C. CURTIN and L. BOX (1986) *The Commoditisation Debate: Labour process, Strategy and Social Network*, Dept. of Rural Sociology, Wageningen, paper no. 17.

MACKENZIE, S. (1989) 'Restructuring the relations of work and life: women as environmental actors, feminism as geographical analysis', in Kobayashi, A. and S. Mackenzie (eds) *Remaking Human Geography*, pp. 40–61. (London: Unwin Hyman).

MACKENZIE, S. and D. ROSE (1983) 'Industrial change, the domestic economy and home life', in J. Anderson, S. Duncan and R. Hudson (eds), *Redundant Spaces in Cities and Regions?* (London: Academic Press) pp. 155–200.

MACKINTOSH, M. (1979) 'Domestic labour and the household', in S. Burman (ed.), *Fit Work for Women* (London: Croom Helm), pp. 173–91.

MACKINTOSH, M. (1981) 'Gender and economics: the sexual division of labour and the subordination of women', in Young, K. *et al.* (ed.), *Of Marriage and the Market*, (London: Routledge & Kegan Paul) pp. 3–17.

MANN, S. and J. DICKENSON (1978), 'Obstacles to the development of capitalist agriculture', *Journal of Peasant Studies*, 5, pp. 466–81.

MARK-LAWSON, J. and A. WITZ (1986) 'From "family labour" to "family wage"', *Lancaster Regionalism Group working paper no. 18*, Lancaster.

MARSDEN, T. K. (1979) 'The socio-economic structure of farming in North Humberside: a study of the farm family in capitalist agriculture', unpublished PhD thesis, University of Hull.

MARSDEN, T. K. (1984) 'Capitalist farming and the farm family', *Sociology*, 18, pp. 205–24.

MARSDEN, T. K. and J. K. LITTLE (eds) (1990) *The British Agro-food System* (Aldershot: Gower).

MARSDEN, T. K., R. J. C. MUNTON, S. J. WHATMORE and J. K. LITTLE (1986a) 'Towards a political economy of capitalist agriculture: a British perspective', *International Journal of Urban and Regional Research*, 10, pp. 498–521.

MARSDEN, T. K., S. J. WHATMORE, R. J. C. MUNTON and J. K. LITTLE (1986b) 'The restructuring process and economic centrality in capitalist agriculture', *Journal of Rural Studies* 2/4, pp. 271–80.

MARSDEN, T. K., S. J. WHATMORE and R. J. C. MUNTON (1987) 'Uneven development and the restructuring process in British agriculture', *Journal or Rural Studies*, 3/4, pp. 297–308.

MARSDEN, T. K. and D. SYMES (1985) 'Intensive livestock units in north Humberside', in B. Ilbery and M. Healy, *The Industrialisation of the Countryside* (Norwich: Geobooks).

MARSHALL, B. (1988) Feminist theory and critical theory. *Canadian Review of Sociology and Anthropology* 25, pp. 208–230.

MARX, K. (1969) *Theories of Surplus Value* (London: Lawrence & Wishart).

MARX, K. (1976) *Capital*, vols 1 and 3 (Harmondsworth: Penguin).

MASSEY, D. (1984) *Spatial Divisions of Labour* (London: Hutchinson).

MASSEY, D. and A. CATALANO (1978) *Capital and Land* (London: Arnold).

MASSEY, D. and R. MEEGAN (eds) (1985) *Politics and Method: Contrasting Studies in Industrial Geography* (London: Methuen).

McDONOUGH, R. and R. HARRISON (1978) 'Patriarchy and relations of production', in A. Kuhn and A. Wolpe, *Feminism and Materialism* (London: Routledge & Kegan Paul), pp. 11–41.

McDOWELL, L. (1986) 'Beyond Patriarchy: a class-based explanation of women's subordination', *Antipode* 18/3, pp. 311–21.

McDOWELL, L. and S. BOWLBY (1983) 'Teaching feminist geography', *Journal of Geography and Higher Education*, 7/2, pp. 97–107.

McINTOSH, M. (1979) 'The welfare state and the needs of the dependent family', in S. Burman (ed.), *Fit Work for Women* (London: Croom Helm).

McMICHAEL, P. and F. BUTTEL (1989) *New directions in the political economy of agriculture*, Cornell University, mimeo.

McROBBIE, A. (1982) 'The politics of feminist research: between talk, text and action', *Feminist Review*, 12, pp. 46–58.

MIES, M. (1982) 'The dynamics of the sexual division of labour and the integration of rural women into the world market', in L. Beneria (ed.), *Women and Development* (New York: Praeger).

MIES, M. (1986) *Indian Women in Subsistence and Agricultural Labour*, (Geneva: International Labour Office) Women, work and development series no. 12.

MINGIONE, E. (1983) 'Informalisation, restructuring and the survival strategies of the working class'. *International Journal of Urban and Regional Research*, 7/3, pp. 311–39.

MITCHELL, C. (1983) 'Case and situation analysis', *The Sociological Review*, 31/2, pp. 187–211.

MITCHELL, J. (1971) *Women's Estate* (Harmondsworth: Penguin).

MOLYNEUX, M. (1977) 'Androcentrism in Marxist anthropology', *Critique of Anthropology*, 9/10, pp. 55–81.

MOLYNEUX, M. (1979) 'Beyond the domestic labour debate', *New Left Review*, 116, pp. 3–28.

MOMSEN, J. and J. TOWNSEND (eds) (1987) *Geography of Gender in the Third World* (London: Hutchinson).

MOONEY, P. (1982) 'Labour time, production time and capitalist development in agriculture: a reconsideration of the Mann–Dickenson thesis, *Sociologia Ruralis*, 22, pp. 279–92,

MOONEY, P. (1983) 'Towards a class analysis of mid-western agriculture', *Rural Sociology*, 48, pp. 563–84.

MOORE, H. (1988) *Feminism and Anthropology* (Cambridge: Polity Press).

MORGAN, D. J. H. (1975) *Social Theory and the Family* (London, Routledge & Kegan Paul).

MOUZELIS, N. (1976) 'Capitalism and the development of agriculture', *Journal of Peasant Studies*, 7, pp. 483–92.

MUNTON, R. J. C. (1983) *London's Green Belt: Containment in Practice* (London: Allen & Unwin).

MUNTON, R. J. C., S. J. WHATMORE and T. K. MARSDEN (1988) 'Reconsidering urban-fringe agriculture. A longitudinal analysis of capital restructuring on farms in the Metropolitan Green Belt', *Transactions of the Institute of British Geographers*, 13/1, pp. 324–36.

MURGATROYD, L. (1985) 'The production of people and domestic labour revisited', in P. Close and R. Collins, *Family and Economy in Modern Society* (London: Macmillan).

MURGATROYD, L., M. SAVAGE, D. SHAPIRO, J. URRY, S. WALBY and A. WARD, with J. MARK-LAWSON (1984) *Localities, Class and Gender* (London: Macmillan).

MURRAY, R. (1977) 'Value and the theory of rent, part 1', *Capital and Class*, 6, pp. 100–22.

NALSON, J. (1968) *Mobility of farm families: a Study of Occupational and Residential Mobility in an Upland Area of England* (Manchester: Manchester University Press).

NEWBY, H. (1980) 'Rural sociology: a trend report', *Current Sociology*, 28, pp. 1–141.

NEWBY, H. (1982) 'Rural sociology and its relevance to the agricultural economist: a review', *Journal of Agricultural Economics*, 33, pp. 125–66.

NEWBY, H. (1985) 'Locality and rurality', *Regional Studies*, 20, pp. 209–16.

NEWBY, H., C. BELL, C. ROSE and P. SAUNDERS (1978) *Property Paternalism and Power* (London: Hutchinson).

NEWBY, H. and P. UTTING (1983) 'Agribusiness in the UK: social and political implications', paper to the British Sociological Association, April, Aberystwyth.

NICHOLSON, L. (1987) 'Feminism and Marxism. Integrating kinship and the economic.' In Banhabib, S. and Cornell D. (eds), *Feminism as Critique*, pp. 16–30. (Cambridge Polity Press).

OAKLEY, A. (1974) *The Sociology of Housework* (London: Martin Robertson).

OAKLEY, A. (1981) 'Interviewing women: a contradiction in terms?' in H. Roberts (ed.), *Doing Feminist Research*, pp. 30–51.

PAHL, R. E. (1984) *Divisions of Labour* (Oxford: Basil Blackwell).

PHILLIPS, A. and B. TAYLOR (1980) 'Sex and skill: notes towards a feminist economics', *Feminist Review*, 6, pp. 79–88.

PLUMWOOD, V. (1989) 'Sex and gender', *Radical Philosophy*, 51, pp. 2–11.

PORTER, M. (1983) *Home, Work and Class Consciousness* (Manchester: Manchester University Press).

POSTER, M. (1978) *Critical Theory of the Family* (London: Pluto Press).

PRINGLE, R. (1988) *Secretaries Talk* (London: Verso).

QVORTRUP, J. (1985) 'Placing children in the division of labour', in P. Close and R. Collins, *Family and Economy in Modern Society* (London: Macmillan), pp. 129–45.

RAPP, R. (1982) 'Family and class in contemporary America: notes toward an understanding of ideology', in B. Thorne and M. Yalom, *Rethinking the Family* (London: Longman), pp. 168–87.

RAPP, R., E. ROSS, and R. BRIDENTHAL (1979) 'Examining family history', *Feminist Studies*, 5/1.

REDCLIFT, N. (1985) 'The contested domain: gender, accumulation and the labour process', in N. Redclift and E. Mingione, *Beyond employment* (Oxford: Basil Blackwell) pp. 92–125.

REDCLIFT, N. and E. MINGIONE (eds) (1985) *Beyond Employment: Household, Gender and Subsistence* (Oxford: Basil Blackwell).

REIMER, B. (1986) 'Women as farm labour', *Rural Sociology*, 51, pp. 143–55.

RHINEHART, N. and P. BARTLETT (1989) 'The persistence of family farms in United States Agriculture', *Sociologia Ruralis*, 29, pp. 203–25.

ROBERTS, H. (ed.) (1981) *Doing Feminist Research* (London: Routledge & Kegan Paul).

ROSTON M. (1983) *Early Neoclassical economics and the economic role of women* (Open University) Social Science Working Paper.

ROWBOTHAM, S. (1972) *Women, Resistence and Revolution* (Harmondsworth: Penguin).

RUBIN, G. (1975) 'Traffic in women: notes on the political economy of sex', in R. Reiter (ed.), *Toward an Anthropology of Women* (New York: Monthly Review Press) pp. 157–210.

SACHS, C. (1983) *Invisible Farmers, Women's work in Agricultural Production* (Totowa, NJ: Rhinehart and Allenheld).

SALAMON, S. and A. M. KEIM (1979) 'Landownership and women's power in a mid-western farming community'. *Journal of Marriage and the Family*, 41, pp. 109–19.

SARRE, P. (1987) 'Realism in practice', *Area*, 19/1, pp. 3–10.

SAYER, A. (1983) 'What kind of Marxism for what kind of feminism?', paper presented at the Women and Geography Study group conference, 'Meeting of Minds', May, Reading.

SAYER, A. (1984) *Method in Social Science. A Realist Approach* (London: Hutchinson).

SCOTT, A. MacEWAN (1986a) 'Why rethink petty commodity production?', Introduction to a special issue of *Social Analysis*, 20, pp. 3–10.

SCOTT, A. MacEWAN (1986b) 'Towards a rethinking of petty commodity production', *Social Analysis*, 20, pp. 93–105.

SCOTT, A. MacEWAN (ed.) (1986) Collection of papers on petty commodity production in a special issue of the journal *Social Analysis*, vol. 20.

SECCOMBE, W. (1986) 'Patriarchy stabilized: the construction of the male breadwinner wage norm in nineteenth-century Britain', *Social History*, 11/11, pp. 53–76.

SHARPE, B. (1988) 'Informal work and development in the west', *Progress in Human Geography*, 9, pp. 315–36.

SIISKONEN, P., A. PARVIAINEN and T. KOPPA (1982) 'Women in agriculture. A study of the position of women engaged in agriculture in Finland in 1980', Pellervo Economic Research Institute discussion paper no. 27.

SINCLAIR, P. (1980) 'Agricultural policy and the decline of commercial family farming: a comparative analysis of US, Sweden and the Netherlands', in F. Buttel and H. Newby, *The Rural Sociology of Advanced Societies* (Montclair, NJ: Allenheld, Osman and Co.) pp. 327–53.

SMITH, C. (1986) 'Forms of production. Fresh approaches to simple commodity production', *Journal of Peasant Studies*, 11, pp. 201–11.

SMITH, C. (1986) 'Reflections on the social relations of simple commodity production', *Journal of Peasant Studies*, 13, pp. 60–95.

SMITH, D. (1983) 'Women, class and family', *Socialist Register* (London: Merlin Press).

SMITH, P. (1978) 'Domestic labour and Marx's theory of value', in A. Kuhn and A. Wolpe, *Feminism and Materialism* (London: Routledge & Kegan Paul) pp. 198–219.

SMITH, S. J. (1981) 'Humanistic methods in contemporary social geography', *Area*, 13/4, pp. 293–8.

SPENDER, D. (1980) *Man-made Language* (London: Routledge & Kegan Paul).

STACEY, M. (1986) 'Gender and stratification. One central issue or two? in R. Compton and N. Mann, *Gender and stratification* (Cambridge: Polity Press) pp. 214–223.

STEBBING, S. (1984) 'Women's roles in rural society', in T. Bradley and P. Lowe (eds), *Locality and Rurality* 199–208.

STINCHCOMBE, A. L. (1961) 'Agricultural enterprise and rural class relations', *American Journal of Sociology*, 67, pp. 165–77.

STIVENS, M. (1981) 'Women, kinship and capitalist development', in K. Young *et al.* (ed.), *Of Marriage and the Market* (London: Routledge & Kegan Paul) pp. 178–92.

STOLKE, V. (1981) 'Women's labours: the naturalisation of social inequality and women's subordination', in K. Young *et al.* (ed.), *Of Marriage and the Market* (London: Routledge & Kegan Paul) pp. 159–77.

STRATEGAKI, M. (1988) 'Modernisation of agricultural production and the gender division of labour: the case of Heraklion, Crete', *Sociologia Ruralis*, 28/4, pp. 248–62.

STRATHERN, M. (1984) 'The social meaning of localism', in T. Bradley and P. Lowe, *Locality and Rurality* (Norwich: Geobooks) pp. 181–98.

SWEET, J. (1972) 'The employment of rural farm wives', *Rural Sociology*, 37/4, pp. 553–77.

SYMES, D. and J. APPLETON (1986) 'Family goals and kinship strategies in a capitalist farming society', *Sociologia Ruralis*, 26, pp. 346–63.

SYMES, D., and T. K. MARSDEN. (1983) 'Complementary roles and asymmetrical lives. Farmers' wives in a large farm environment', *Sociologia Ruralis*, 23/3–4, pp. 229–41.

SZELENYI, I. (1988) *Socialist Entrepreneurs: Embourgeoisement in Rural Hungary* (Cambridge: Polity Press).

THOMAS, D. (1985) *Citizenship, Gender and Work: Social Organisation of Industrial Society* (Berkeley: University of Berkeley, California).

THOMPSON, E. P. (1967) 'Time, work-discipline and industrial capitalism', *Past and Present*, 37/38, pp. 56–97.

THORNE, B. and M. YALOM (eds) (1982) *Rethinking the Family. Some Feminist Questions* (London: Longman).

THRIFT, N. (1983) 'On the determination of social action in space and time', *Society and Space*, 1/1, pp. 23–57.

THRIFT, N. (1987) 'No perfect symmetry. A response to David Harvey', *Society and Space*, 5, pp. 400–7.

TRAILL, W. B. (1980) 'Land values and rents' (Department of Agricultural Economics, University of Manchester) *Bulletin no. 175*.

URRY, J. (1981) *The Anatomy of Capitalist Societies: the Economy, Civil Society and the State* (London: Macmillan).

URRY, J. (1984) 'Capitalist restructuring, recomposition and the regions', in T. Bradley and P. Lowe (eds.), *Locality and Rurality* (Norwich: Geobooks) pp. 45–64.

VAN DER PLOEG, J. D. (1986) 'The agricultural labour process and commoditisation', in N. Long *et al.*, (eds), *The Commoditisation Debate* pp. 24–57.

VERGOPOLOS, K. (1978) 'Capitalism and peasant productivity', *Journal of Peasant Studies*, 5, pp. 446–514.

VOGEL, L. (1983) *Marxism and the Oppression of Women. Towards a Unitary Theory* (London: Pluto).

WALBY, S. (1986) *Patriarchy at Work* (Cambridge: Polity Press).

WALBY, S. (1989) 'Theorising patriarchy', *Sociology*, 23/2, pp. 213–34.

WALLMAN, S. (1984) *Eight London Households* (London: Tavistock Press).

WEEDON, C. (1987) *Feminist Practice and Poststructuralist Theory* (Oxford: Basil Blackwell).

WEEKS, J. (1985) *Sexuality and its Discontents: Meanings, Myths and Modern Sexualities* (London: Routledge & Kegan Paul).

WHATMORE, S. J. (1986) 'Landownership and the development of modern British agriculture', in G. Cox., P. Lowe and M. Winter, *Agriculture: People and Policies* (London: Allen & Unwin) pp. 105–25.

WHATMORE, S. J. (1988) 'From women's roles to gender relations. Changing perspectives in the analysis of farm women', Introduction to a special issue of *Sociologia Ruralis*, on farm women in Europe, 28, pp. 239–47.

WHATMORE, S. J., R. J. C. MUNTON, T. K. MARSDEN and J. K. LITTLE (1987a) 'Towards a typology of farm businesses in contemporary British agriculture', *Sociologia Ruralis* 27/1, pp. 21–37.

WHATMORE, S. J., R. J. C. MUNTON, T. K. MARSDEN and J. K. LITTLE (1987b) 'Interpreting a relational typology of farm businesses in southern England', *Sociologia Ruralis*, 27/2, pp. 103–22.

WHITEHEAD, A. (1981) ' "I'm hungry, Mum": the politics of domestic budgeting', in K. Young *et al.*, *Of Marriage and the Market*, (London: Routledge & Kegan Paul) pp. 93–116.

WHITEHEAD, A. (1984) 'Men, and women, kinship and property: some general issues', in R. Hirschon (ed), *Women and Property, Women as property* (London: Croom Helm) pp. 176–210.

WHYTE, W. F. (1955) *Street Corner Society* (Chicago: University of Chicago Press).

WILLIAMS, R. (1983) *Towards 2000* (Harmondsworth: Penguin).

WILLIAMS, R. (1984) 'Between the country and the city', in R. Mabey, S. Clifford and A. King, (eds) *Second Nature* (London: Jonathan Cape).

WILLIAMS, W. M. (1964) *A West Country Village. Ashworthy* (London: Routledge & Kegan Paul).

WILSON, F. (1984) 'Women and the commercialisation of agriculture: a review of recent literature on Latin America, mimeo (Copenhagen).

WINTER, M. (1982) 'Whatever happened to the agrarian bourgeoisie and the rural proletariat under monopoly capitalism? A reply to Goran Djurfeldt', *Acta Sociologica*, 25/2, pp. 147–57.

WINTER, M. (1984) 'Agrarian class structure and family farming', in T. Bradley and P. Lowe (eds), *Locality and Rurality* (Norwich: Geobooks) pp. 115–28.

WINTER, M. (1986) 'The development of family farming in west Devon in the nineteenth century', in G. Cox *et al.* (ed) *Agriculture: People and Policies* (London: Allen & Unwin) pp. 61–76.

WOLPE, H. (ed) (1980) *The Articulation of Modes of Production* (London: Routledge & Kegan Paul).

WOMEN AND GEOGRAPHY STUDY GROUP (1984) *Gender and Geography* (London: Hutchinson).

WRIGHT, E. (1989) 'Women in the class structure', *Politics and Society*, 17, pp. 35–66.

WRIGHT, P. (1985) *On Living in an Old Country* (London: Verso).

YANAGISAKO, S. (1979) 'Family and household: the analysis of domestic groups', *Annual Review of Anthropology*, 8, pp. 161–205.

YOUNG, K., C. WALKOWITZ and R. McCULLAGH (eds) (1981) *Of Marriage and the Market* (London: Routledge & Kegan Paul).

ZARETSKY, E. (1976) *Capitalism, the Family and Personal Life* (London: Pluto).

Index

['n' indicates a citation in the notes. **bold** indicates a definition of terms]

Acock, A., 164n
Albert, C., 158n
Agriculture Economic
    Development Committee, 5
agriculture, restructuring of, 3–4,
    16–17, 18, 27, 144
agroindustrial complex, 17–18, 53
Allatt, P., 42
Allen, J., 160n
Allen, S., 8
Andre, J., 35
Anthias, F., 165n
Antipode, 161n
Appleton, J., 65, 162n
Ardener, E., 47
Ardener, S., 3, 8
Assiter, A., 162n
Ayim, M., 35

Ball, M., 16, 159n
Banajii, J., 13
Barker, D., 8
Barkley, P., 159n
Barlett, P., 23
Barlow, J., 159n
Barrett, M., 29, 40, 41, 160n
Bartells, A., 35
Barthez, A., 158n
Bauwens, A., 158n
Bechhoffer, F., 2, 145
Beechey, V., 160n, 165n
Beneria, L., 37, 39, 46, 158n
Bennholt-Thompson, V., 38
Benvenuti, B., 2, 144
Berger, J., 162n
Berk, S., 46, 47, 163n
Bernades, J., 40
Bernstein, H., 2, 21, 24, 160n
Bhaskar, R., 162n
Bland, L., 37

Boserup, E., 158n
Boulding, E., 158n
Bouquet, M., 9, 13, 65, 69, 72, 73,
    83, 162n, 164n
Bourdieu, P., 7, 48, 161n, 162n,
    165
Bowlby, S., 29, 48, 142
Bradby, B., 29, 34, 39
Bradley, T., 12, 159n
Brass, T., 33
Breughal, I., 165n
Bridenthal, R., 22
Brittan, A., 161n
Bromley, R., 1
Bryceson, D., 161n
Buchanan, W., 163n, 164n
Burawoy, M., 27, 43, 46, 50, 86,
    103, 158n
Burgess, J., 48
Burns, J., 159n
Burrell, A., 4
Buttel, F., 12, 159n, 164n

Caldwell, L., 160n
Callen, H., 3, 8
Canoves, G., 159n
Carvacao, C., 159n
case studies, *see* methodology
Catalano, A., 159n
CCCS, 48
Chayanov, V., 13
Chaytor, M., 160n, 164n
Chevalier, J., 22
Cicourel, A., 162n
Close, P., 2
Clutterbuck, C., 159n
Cochrane, W., 18
Cockburn, C., 3, 104, 145–6, 162n,
    166n
Collins, R., 2, 41, 66

184

Comer, L., 41
commoditisation, **6**, 27, 33, 143
  debate, 12 (*see also* constraints
    thesis, resilience thesis)
  different levels of, 25, 54–5, 65,
    83, 94
conjugal family household, *see*
  household
Connell, R., 34–7, 40, 50, 161n
constraints thesis, 13, 15–19, 140
Conway, J., 13, 159n
Cooke, P., 159n
Coulson, M., 28
Coward, R., 35
Creighton, C., 162n
Crompton, R., 145, 160n
Crow, G., 6
cumulative interviewing, 60–2
Curtin, C., 31

Davidoff, L., 2, 9, 104
Davis, J., 13
decision-making, 80–1
Delphy, C., 34, 35, 39, 103, 161n,
  166n
Deseran, F., 164n
Dickenson, J., 16
division of labour
  by gender, 5, 33, 44, 56, 73, 83,
    99
  family/household, 31–2, 44, 66–7
Dixon-Mueller, R., 46, 63, 164n
domestic commodity production **31**,
  31–3, 82–4, 140, 143
domestic labour
  debate, 28 (*see also* feminism)
  see also farm labour process
domestic political economy, 9, 43–
  5, 105–6, 140
Donzelot, J., 37
Dorset (west), 51, 52–3, 54–7, 67, 72
Duncan, S., 159n

Edholm, F., 21, 31, 38
Eisenstein, H., 36
Elliott, B., 2, 145
Ennew, J., 159n
Equal Opportunities Commission,
  47, 162n, 164n

Errington, A., 4, 57
ethnography, *see* methodology
'everyday sense-making', 50, 102–3
external relations (of farm
  production), 14, 18, 53–5, 105
Eyles, J., 162n

family, 'the', 6, 28, 31, 33, 40–3, 146
  enterprise, 43, 145
  labour process, 27, 33, 44, 141–2
  see also household; kinship
family farm, 4, **6**, 14, 19, 20, 54,
  143
  family farming, 12, 139–40
farm labour process, 5, 7–8, 27,
  43–4, 141
  domestic household labour, 44,
    66–8, 82–3, 87, 103
  agricultural labour, 44, 68–71,
    83, 109, 113–14, 120
  non-agricultural farm labour, 45,
    71–2, 115, 124–6, 131
  off-farm wage labour, 45, 72,
    131–2
farm women, 4, 60, 143–4
  as wives, 5, 63, 66–9, 86–9, 94
    (*see also* wifehood)
  survey of, *see* methodology; case
    studies
feminism, 2–3, 29
  and methodology, 46–9
  feminist theory, 29, 34–8 (*see
    also* patriarchy, gender
    relations)
  Marxist feminist, 28 (*see also*
    domestic labour debate)
  relationship to Marxism, 147–8
femininity, 34, 142
  see also wifehood
Finch, J., 3
Fine, B., 16
Fitzsimmons, M., 160n
Flax, J., 40
Flora, C. Butler, 83
Folbre, N., 6
Foord, J., 7, 35– 7, 161n
Friedland, W., 18, 159n
Friedmann, H., 12, 13, 14, 19–24,
  28, 30–3, 65, 160n, 161n

Gamarnikow, E., 44
Game, A., 162n
Garcia, M. Dolors, 159n
Gardiner, J., 160n
Garnsey, E., 29
Gasson, R., 5, 6, 69, 70, 71, 158n, 164n, 165n
Geertz, C., 48–9
Geisler, C., 74
gender, 32, 144
  identities, 34, 40, 59, 103, 142
  ideologies, 8, 43, 85, 142 (*see also* wifehood)
  order, 37, 40, 142, 146
  regime, 37, 40, 45, 104, 142
  relations, 7, 35
    theory of, 34–7 (*see also* role theory)
  see also patriarchy, division of labour
generation, 32
Gerry, C., 1
Ghorayshi, P., 5, 159n
Giddens, A., 162n
Gillespie, G., 164n
Gittins, D., 163n
Goodman, D., 12, 13, 14, 15–19, 16, 21, 24, 159n
Goss, K., 17, 159n
Graham, H., 48
Gray, A., 160n
Gregson, N., 35–7, 161n, 163n
Grigg, D., 4

Hall, C., 2, 104
Hall, S., 49, 162n
Haney, W., 158n
Harré, R., 160n
Harris, C., 28, 37, 41, 160n
Harris, O, 31, 160n
Harrison, A., 4
Harrison, J., 160n
Harrison, R., 28
Hart, N., 145
Hartmann, H., 34, 35, 39
Harvey, D., 159n
Healey, M., 159n
Hearn, J., 161n
Hedley, M., 31, 41

Heller, A., 50, 162n
Himmelweit, S., 21
Hirschon, R., 41, 73
Holstrom, N., 161n
household, 33, 40–3, 41
  conjugal, 8, 40–3, 42, 73–4, 141
  economics, 29
  see also family, marriage
Houston, B., 35
human agency, 27–7, 106, 147
Humphries, J., 158n
Hunt, P., 165n
Hussein, A., 12
Hutson, J., 162

Ilbery, B., 159n
internal relations (of farm production), 14, 18, 32, 53–5, 84, 140

Janvry, A. de, 13, 19, 159n
Jones, C., 70, 158n
Jones, J., 159n

Kahn, J., 159n
Kaluzynska, E., 160n
Kaplan, C., 147–8, 162n
Kautsky, K., 13
Kazi, H., 3
Keat, R., 162n
Keim, A., 74
Kenny, M., 158n
kinship (kin), 40–2, 74, 84, 141
  see also family, marriage
Kloppenburg, J., 17, 158n
Knorr-Cetina, K., 49, 50, 162n
Knowles, B., 158n
Kuhn, A., 28, 48, 160n

labour circuits, 44, 65, 83, 141
  see also farm labour process
labour process, *see under* family, farm labour process
Lagrave, R., 158n
Lang, T., 159n
Laslett, B., 163n
Lenin, V., 159n
Leonard, D., 39
Leonardo, M. di, 162n
life-course, 42

life-cycle, 42, 65, 82
Little, J., 159n
Littlejohn, J., 4
livelihood, 1, 39, 139, 147
  see also subsistence
Loeffen, T., 158n
Long, N., 2, 12, 13, 15, 20, 24–5,
  26, 46–7
Lowe, P., 159n
Luckmann, T., 162n

Mackenzie, S., 147, 165n
Mackintosh, M., 32, 41, 161n
Mann, M., 145, 160n
Mann, S., 16
Mark-Lawon, J., 158n
marriage, 8, 40–1, 84, 100
  see also kinship, household,
  wives
Marsden, T., 24, 25, 42, 47, 65, 79,
  159n, 160n
Marshall, B., 147
Marx, K., 16, 30
Marxism, structuralist, 7, 13
  see also political
  economy, feminism
masculinity, 34, 133
Massey, D., 7, 159n, 163n
McDonough, R., 28
McDowell, L., 35, 48
McIntosh, M., 8, 41, 160n
McRobbie, A., 29, 49
Meegan, R., 163n
men, 5, 34, 83, 100, 104
methodology, 46–9
  base-line farm survey, 53
  case studies, 51, 59–64
  ethnography, 51, 61–2
  survey of farm wives, 50, 55–9
  see also feminism, typology
Metropolitan Green Belt, 51–2,
  53–8, 72
Mies, M., 29, 158n, 164n
Mingione, E., 1, 147
Mitchell, C., 63
Mitchell, J., 35
Mohun, J., 21
Molyneux, M., 5, 160n, 161n
Momsen, J., 158n

Mooney, P., 16, 159n
Moore, H., 3, 29, 158n, 162n
Morgan, D., 160n
Mouzelis, N., 159n
Munton, R., 51, 52
Murgatroyd, L., 44, 46, 159n, 160n
Murray, R., 16

Nalson, J., 79
Newby, H., 6, 9, 18, 159n
Nicholson, L., 29
Nix, J., 74

Oakley, A., 46, 162n

Pahl, R., 44, 162n
patriarchy, 2, 29, 35–6, 42, 145–6
  patriarchal gender relations, 29,
  34–7, 43, 141, 144
  relationship to capitalism, 36–7,
  59, 146
  see also gender, feminism
petty commodity production, 1,
  12, 30, 143
Phillips, A., 162n
Plumwood, V., 161n
political economy, 12, 139, 148
  agrarian, **12**, 27
  Marxist, 2, 6, **7**, 27, 30, 32, 34
Poster, M., 160n
post-structuralism, 7, 24, 147–8
  in political economy analysis,
  25–7, 43, 139
  see also human agency
Pringle, A., 3, 162n
production, relationship to
  reproduction, 38–9, 94, 105,
  141, 146
  see also domestic political
  economy
property
  in family enterprise, 4, 19, 21,
  33, 54, 73–4, 107, 112–13, 118,
  123, 133–4, 142
  women's, 74–80, 84, 100, 118,
  144
  see also kinship

Qvortrup, J., 162n, 164n

Rapp, R., 28, 40, 161n
Rappaport, R., 163n
Redclift, M., 12, 13, 14, 15–19, 21, 24, 159n
Redclift, N., 1, 3, 29, 38, 39, 147
Reimer, B., 5
Reinhart, N., 23
reproduction, process of, 20, 21–3, 30–1, 42, 76, 108, 128
  biological reproduction, 36–9
  expanded reproduction, 22–3, 38
  reproduction of labour power, 39, 66
resilience thesis, 13, 19–24, 140
Roberts, E., 47, 162n
role theory, 35
Rose, D., 165n
Rosenfeld, R., 70, 158n
Roston, M., 161n
Rowbotham, S., 2
Rubin, G., 34
Rural Sociologist, 158n

Sachs, C., 9, 158n
Salamon, S., 74
Sarre, P., 163n
Savage, M., 159n
Sayer, A., 27, 50, 63, 139, 147, 161n, 162n, 163n
Scott, A. MacEwan, 15, 24, 31, 32, 160n
Seccombe, W., 103
self-exploitation, 23–4
sex, 34
  sex roles, 35 (*see also* role theory)
  sexuality, 40, 41, 103, 142
Sharpe, B., 162n
Shih, A., 47
Siiskonen, P., 158n
simple commodity production, 19–20, 20–4, 30–3, 143
Sinclair, P., 13, 19
Smith, C., 24, 31
Smith, D., 3
Smith, David, 162n
Smith, P., 21
Smith, S., 49
Social Analysis, 159n

social practice, 49–50
Sociologia Ruralis, 14, 158n
spatial imagery, 2–3, 9, 39, 145–6
spatial relations
  of the gender division of labour, 95, 97, 100
  of production and reproduction, 38–9, 97, 100, 146–7
Spender, D., 162n
Stacey, M., 146
Stebbing, S., 162n
Stinchcombe, A., 159n
Stivens, M., 42
Stolke, V., 34
Strategaki, M., 42
Strathern, M., 48, 142
subsistence, 39, 94–5
  see also livelihood
subsumption, 15–16, 19
Sweet, J., 164n
Symes, D., 47, 65, 159n, 162n
Szelenyi, I., 158n

Taylor, B., 162n
Thomas, R., 159n
Thompson, E., 46
Thorne, B., 160
Thrift, N., 7, 27
time diary, 63
Townsend, J., 158n
Traill, W., 74
Tribe, K., 12
typology, of farm businesses, 25–6, 53–5, 105–6
  family business farm, 55, 57–9, 111–17
  family labour farm, 55, 57–9, 107–11
  transitional family farm, 55, 57–9, 117–27

Urry, J., 2, 162n
Utting, P., 18

van der Ploeg, J., 18, 22–3, 24, 31
Vergopolos, K., 159n
Vogel, L., 35

Walby, S., 35, 161n, 166n

Wallman, C., 48, 86, 165n
Weedon, S., 48, 86, 162n
Weeks, J., 161n
Whatmore, S., 16, 26, 54
Whitehead, A., 41, 160n
Whyte, W., 61
wifehood, 10, 59, 84–5, 86–7,
   92–4, 103
   see also gender ideologies
Williams, R., 2, 144, 146–7, 161n
Williams, W. M., 4
Wilson, F., 158n
Winter, M., 13, 14, 15
Witz, A., 159n
Wolpe, A., 28, 48, 161n
Women and Geography Study
   Group, 2

women, 5, 34
   as wives, 5, 8, 33, 43, 142 (*see
      also* farm women)
work
   ideologies of, 27, 43, 103
   non-wage forms, 1, 46–7, 92–3,
      95, 145
   subjective experience/meaning of,
      26–7, 90–2, 102
Wright, P., 50

Yalom, M., 160n
Yanagisako, S., 41, 162n
Young, K., 8

Zaretsky, E., 28